I0038653

Introduction to Geographic Information Systems

Introduction to Geographic Information Systems

Edited by
Jayceon Garcia

Larsen & Keller
www.larsen-keller.com

Introduction to Geographic Information Systems
Edited by Jayceon Garcia
ISBN: 978-1-63549-133-3 (Hardback)

Published by Larsen and Keller Education,
5 Penn Plaza,
19th Floor,
New York, NY 10001, USA

Cataloging-in-Publication Data

Introduction to geographic information systems / edited by Jayceon Garcia.
p. cm.
Includes bibliographical references and index.
ISBN 978-1-63549-133-3
1. Geographic information systems. 2. Information storage and retrieval systems--Geography.
I. Garcia, Jayceon.
G70.212 .I58 2017
910.285--dc23

Printed and bound in the United States of America.

For more information regarding Larsen and Keller Education and its products, please visit the publisher's website www.larsen-keller.com

Table of Contents

Preface

The book aims to shed light on some of the unexplored aspects of geographic information systems (GIS). It provides thorough knowledge about the basic principles of this subject. Geographic information systems refer to a system that is used to manipulate, capture, store, analyze and maintain geographical data. The textbook covers the various fields that use this technology as well as the methods associated with it. It also focuses on the theoretical aspects on geographic information science. For someone with an interest and eye for detail, this book covers the most significant topics in this field. It will serve as a valuable source of reference for those interested in GIS.

A short introduction to every chapter is written below to provide an overview of the content of the book:

Chapter 1 - Geographical data needs to be analyzed, managed and then collected. The process by which this is done is known as a geographic information systems. This chapter also focuses on topics such as geographic information science, spatial data infrastructure and spatial database. The text is an overview of the subject matter incorporating all the major aspects of geographic information systems; **Chapter 2** - The major elements of geographic information systems are GIS file formats, conservation geoportal, data model, digital mapping, distributed GIS etc. A GIS file format is the conversion of geographical information into a computer file. Geographical information systems need a complete listing of data on an online portal. The online portal for these data sets is known as conservation geoportal. The topics discussed in the chapter are of great importance to broaden the existing knowledge on geographic information systems; **Chapter 3** - Tools and techniques are an important component of any field of study. The tools and techniques discussed within this content are ichthyology and GIS, spatial analysis, geotagged photography, remote sensing and rubbersheeting. A GIS can provide accurate data of underwater geography whereas spatial analysis is the technique used to study geographical properties. The following chapter helps the reader in developing an in-depth understanding of the techniques used in geographic information systems; **Chapter 4** - Geocoding is the practice of converting a location to an address on the Earth's surface. The opposite of geocoding is reverse geocoding; where the process is of back coding a location to a readable address. This chapter is an overview of the subject matter incorporating all the major aspects of geocoding; **Chapter 5** - The applications of geographic information systems explained in the chapter are GIS applications, satellite imagery, crime mapping, Google Earth, map algebra, map regression etc. The software and hardware systems that enable users to capture, store and manage geographic data are known as geographic information systems. The text strategically encompasses and incorporates the major application of geographic information systems; **Chapter 6** - A geographic information systems are an interdisciplinary subject. This section will provide a glimpse of the related fields of geographic information systems. GIS and public health, GIS and aquatic science, GIS and hydrology and GIS and environmental governance are some of the aspects elucidated in the text; **Chapter 7** - The geographic information systems that stores and analyses data of the past geographies is referred to as historical geographic information systems. The following content mentions topics such as Great Britain historical GIS and HistoAtlas. The chapter serves as a source to understand the historical perspective of geographic information systems.

I extend my sincere thanks to the publisher for considering me worthy of this task. Finally, I thank my family for being a source of support and help.

Editor

1

Introduction to Geographic Information Systems

Geographical data needs to be analyzed, managed and then collected. The process by which this is done is known as a geographic information system. This chapter also focuses on topics such as geographic information science, spatial data infrastructure and spatial database. The text is an overview of the subject matter incorporating all the major aspects of geographic information system.

Geographic Information System

A geographic information system or geographical information system (GIS) is a system designed to capture, store, manipulate, analyze, manage, and present all types of spatial or geographical data. The acronym GIS is sometimes used for geographic information science (GIScience) to refer to the academic discipline that studies geographic information systems and is a large domain within the broader academic discipline of geoinformatics. What goes beyond a GIS is a spatial data infrastructure, a concept that has no such restrictive boundaries.

In a general sense, the term describes any information system that integrates, stores, edits, analyzes, shares, and displays geographic information. GIS applications are tools that allow users to create interactive queries (user-created searches), analyze spatial information, edit data in maps, and present the results of all these operations. Geographic information science is the science underlying geographic concepts, applications, and systems.

GIS is a broad term that can refer to a number of different technologies, processes, and methods. It is attached to many operations and has many applications related to engineering, planning, management, transport/logistics, insurance, telecommunications, and business. For that reason, GIS and location intelligence applications can be the foundation for many location-enabled services that rely on analysis and visualization.

GIS can relate unrelated information by using location as the key index variable. Locations or extents in the Earth space–time may be recorded as dates/times of occurrence, and x, y, and z coordinates representing, longitude, latitude, and elevation, respectively. All Earth-based spatial–temporal location and extent references should, ideally, be relatable to one another and ultimately to a "real" physical location or extent. This key characteristic of GIS has begun to open new avenues of scientific inquiry.

History of Development

The first known use of the term "geographic information system" was by Roger Tomlinson in the year 1968 in his paper "A Geographic Information System for Regional Planning". Tomlinson is also acknowledged as the "father of GIS".

E. W. Gilbert's version (1958) of John Snow's 1855 map of the Soho cholera outbreak showing the clusters of cholera cases in the London epidemic of 1854

Previously, one of the first applications of spatial analysis in epidemiology is the 1832 "*Rapport sur la marche et les effets du choléra dans Paris et le département de la Seine*". The French geographer Charles Picquet represented the 48 districts of the city of Paris by halftone color gradient according to the number of deaths by cholera per 1,000 inhabitants. In 1854 John Snow determined the source of a cholera outbreak in London by marking points on a map depicting where the cholera victims lived, and connecting the cluster that he found with a nearby water source. This was one of the earliest successful uses of a geographic methodology in epidemiology. While the basic elements of topography and theme existed previously in cartography, the John Snow map was unique, using cartographic methods not only to depict but also to analyze clusters of geographically dependent phenomena.

The early 20th century saw the development of photozincography, which allowed maps to be split into layers, for example one layer for vegetation and another for water. This was particularly used for printing contours – drawing these was a labour-intensive task but having them on a separate layer meant they could be worked on without the other layers to confuse the draughtsman. This work was originally drawn on glass plates but later plastic film was introduced, with the advantages of being lighter, using less storage space and being less brittle, among others. When all the layers were finished, they were combined into one image using a large process camera. Once color printing came in, the layers idea was also used for creating separate printing plates for each color. While the use of layers much later became one of the main typical features of a contemporary

GIS, the photographic process just described is not considered to be a GIS in itself – as the maps were just images with no database to link them to.

Computer hardware development spurred by nuclear weapon research led to general-purpose computer "mapping" applications by the early 1960s.

The year 1960 saw the development of the world's first true operational GIS in Ottawa, Ontario, Canada by the federal Department of Forestry and Rural Development. Developed by Dr. Roger Tomlinson, it was called the Canada Geographic Information System (CGIS) and was used to store, analyze, and manipulate data collected for the Canada Land Inventory – an effort to determine the land capability for rural Canada by mapping information about soils, agriculture, recreation, wildlife, waterfowl, forestry and land use at a scale of 1:50,000. A rating classification factor was also added to permit analysis.

CGIS was an improvement over "computer mapping" applications as it provided capabilities for overlay, measurement, and digitizing/scanning. It supported a national coordinate system that spanned the continent, coded lines as arcs having a true embedded topology and it stored the attribute and locational information in separate files. As a result of this, Tomlinson has become known as the "father of GIS", particularly for his use of overlays in promoting the spatial analysis of convergent geographic data.

CGIS lasted into the 1990s and built a large digital land resource database in Canada. It was developed as a mainframe-based system in support of federal and provincial resource planning and management. Its strength was continent-wide analysis of complex datasets. The CGIS was never available commercially.

In 1964 Howard T. Fisher formed the Laboratory for Computer Graphics and Spatial Analysis at the Harvard Graduate School of Design (LCGSA 1965–1991), where a number of important theoretical concepts in spatial data handling were developed, and which by the 1970s had distributed seminal software code and systems, such as SYMAP, GRID, and ODYSSEY – that served as sources for subsequent commercial development—to universities, research centers and corporations worldwide.

By the late 1970s two public domain GIS systems (MOSS and GRASS GIS) were in development, and by the early 1980s, M&S Computing (later Intergraph) along with Bentley Systems Incorporated for the CAD platform, Environmental Systems Research Institute (ESRI), CARIS (Computer Aided Resource Information System), MapInfo Corporation and ERDAS (Earth Resource Data Analysis System) emerged as commercial vendors of GIS software, successfully incorporating many of the CGIS features, combining the first generation approach to separation of spatial and attribute information with a second generation approach to organizing attribute data into database structures.

In 1986, Mapping Display and Analysis System (MIDAS), the first desktop GIS product emerged for the DOS operating system. This was renamed in 1990 to MapInfo for Windows when it was ported to the Microsoft Windows platform. This began the process of moving GIS from the research department into the business environment.

By the end of the 20th century, the rapid growth in various systems had been consolidated and standardized on relatively few platforms and users were beginning to explore viewing GIS data over the Internet, requiring data format and transfer standards. More recently, a growing number of free, open-source GIS packages run on a range of operating systems and can be customized to perform specific tasks. Increasingly geospatial data and mapping applications are being made available via the world wide web.

GIS Techniques and Technology

Modern GIS technologies use digital information, for which various digitized data creation methods are used. The most common method of data creation is digitization, where a hard copy map or survey plan is transferred into a digital medium through the use of a CAD program, and geo-referencing capabilities. With the wide availability of ortho-rectified imagery (from satellites, aircraft, Helikites and UAVs), heads-up digitizing is becoming the main avenue through which geographic data is extracted. Heads-up digitizing involves the tracing of geographic data directly on top of the aerial imagery instead of by the traditional method of tracing the geographic form on a separate digitizing tablet (heads-down digitizing).

Relating Information from Different Sources

GIS uses spatio-temporal (space-time) location as the key index variable for all other information. Just as a relational database containing text or numbers can relate many different tables using common key index variables, GIS can relate otherwise unrelated information by using location as the key index variable. The key is the location and/or extent in space-time.

Any variable that can be located spatially, and increasingly also temporally, can be referenced using a GIS. Locations or extents in Earth space–time may be recorded as dates/times of occurrence, and x, y, and z coordinates representing, longitude, latitude, and elevation, respectively. These GIS coordinates may represent other quantified systems of temporo-spatial reference (for example, film frame number, stream gage station, highway mile-marker, surveyor benchmark, building address, street intersection, entrance gate, water depth sounding, POS or CAD drawing origin/units). Units applied to recorded temporal-spatial data can vary widely (even when using exactly the same data,), but all Earth-based spatial–temporal location and extent references should, ideally, be relatable to one another and ultimately to a "real" physical location or extent in space–time.

Related by accurate spatial information, an incredible variety of real-world and projected past or future data can be analyzed, interpreted and represented. This key characteristic of GIS has begun to open new avenues of scientific inquiry into behaviors and patterns of real-world information that previously had not been systematically correlated.

GIS Uncertainties

GIS accuracy depends upon source data, and how it is encoded to be data referenced. Land surveyors have been able to provide a high level of positional accuracy utilizing the GPS-derived positions. High-resolution digital terrain and aerial imagery, powerful computers and Web technology are changing the quality, utility, and expectations of GIS to serve society on a grand scale, but nevertheless there are other source data that affect overall GIS accuracy like paper maps, though these may be of limited use in achieving the desired accuracy.

In developing a digital topographic database for a GIS, topographical maps are the main source, and aerial photography and satellite imagery are extra sources for collecting data and identifying attributes which can be mapped in layers over a location facsimile of scale. The scale of a map and geographical rendering area representation type are very important aspects since the information content depends mainly on the scale set and resulting locatability of the map's representations. In order to digitize a map, the map has to be checked within theoretical dimensions, then scanned into a raster format, and resulting raster data has to be given a theoretical dimension by a rubber sheeting/warping technology process.

A quantitative analysis of maps brings accuracy issues into focus. The electronic and other equipment used to make measurements for GIS is far more precise than the machines of conventional map analysis. All geographical data are inherently inaccurate, and these inaccuracies will propagate through GIS operations in ways that are difficult to predict.

Data Representation

GIS data represents real objects (such as roads, land use, elevation, trees, waterways, etc.) with digital data determining the mix. Real objects can be divided into two abstractions: discrete objects (e.g., a house) and continuous fields (such as rainfall amount, or elevations). Traditionally, there are two broad methods used to store data in a GIS for both kinds of abstractions mapping references: raster images and vector. Points, lines, and polygons are the stuff of mapped location attribute references. A new hybrid method of storing data is that of identifying point clouds, which combine three-dimensional points with RGB information at each point, returning a "3D color image". GIS thematic maps then are becoming more and more realistically visually descriptive of what they set out to show or determine.

Data Capture

Example of hardware for mapping (GPS and laser rangefinder) and data collection (rugged computer). The current trend for geographical information system (GIS) is that accurate mapping and data analysis are completed while in the field. Depicted hardware (field-map technology) is used mainly for forest inventories, monitoring and mapping.

Data capture—entering information into the system—consumes much of the time of GIS practitioners. There are a variety of methods used to enter data into a GIS where it is stored in a digital format.

Existing data printed on paper or PET film maps can be digitized or scanned to produce digital data. A digitizer produces vector data as an operator traces points, lines, and polygon boundaries from a map. Scanning a map results in raster data that could be further processed to produce vector data.

Survey data can be directly entered into a GIS from digital data collection systems on survey instruments using a technique called coordinate geometry (COGO). Positions from a global navigation satellite system (GNSS) like Global Positioning System can also be collected and then imported into a GIS. A current trend in data collection gives users the ability to utilize field computers with the ability to edit live data using wireless connections or disconnected editing sessions. This has been enhanced by the availability of low-cost mapping-grade GPS units with decimeter accuracy in real time. This eliminates the need to post process, import, and update the data in the office after fieldwork has been collected. This includes the ability to incorporate positions collected using a laser rangefinder. New technologies also allow users to create maps as well as analysis directly in the field, making projects more efficient and mapping more accurate.

Remotely sensed data also plays an important role in data collection and consist of sensors attached to a platform. Sensors include cameras, digital scanners and lidar, while platforms usually consist of aircraft and satellites. In England in the mid 1990s, hybrid kite/balloons called Helikites first pioneered the use of compact airborne digital cameras as airborne Geo-Information Systems. Aircraft measurement software, accurate to 0.4 mm was used to link the photographs and measure the ground. Helikites are inexpensive and gather more accurate data than aircraft. Helikites can be used over roads, railways and towns where UAVs are banned.

Recently with the development of miniature UAVs, aerial data collection is becoming possible with them. For example, the Aeryon Scout was used to map a 50-acre area with a Ground sample distance of 1 inch (2.54 cm) in only 12 minutes.

The majority of digital data currently comes from photo interpretation of aerial photographs. Soft-copy workstations are used to digitize features directly from stereo pairs of digital photographs. These systems allow data to be captured in two and three dimensions, with elevations measured directly from a stereo pair using principles of photogrammetry. Analog aerial photos must be scanned before being entered into a soft-copy system, for high-quality digital cameras this step is skipped.

Satellite remote sensing provides another important source of spatial data. Here satellites use different sensor packages to passively measure the reflectance from parts of the electromagnetic spectrum or radio waves that were sent out from an active sensor such as radar. Remote sensing collects raster data that can be further processed using different bands to identify objects and classes of interest, such as land cover.

When data is captured, the user should consider if the data should be captured with either a relative accuracy or absolute accuracy, since this could not only influence how information will be interpreted but also the cost of data capture.

After entering data into a GIS, the data usually requires editing, to remove errors, or further processing. For vector data it must be made "topologically correct" before it can be used for some advanced analysis. For example, in a road network, lines must connect with nodes at an intersection. Errors such as undershoots and overshoots must also be removed. For scanned maps, blemishes on the source map may need to be removed from the resulting raster. For example, a fleck of dirt might connect two lines that should not be connected.

Raster-to-Vector Translation

Data restructuring can be performed by a GIS to convert data into different formats. For example, a GIS may be used to convert a satellite image map to a vector structure by generating lines around all cells with the same classification, while determining the cell spatial relationships, such as adjacency or inclusion.

More advanced data processing can occur with image processing, a technique developed in the late 1960s by NASA and the private sector to provide contrast enhancement, false color rendering and a variety of other techniques including use of two dimensional Fourier transforms. Since digital data is collected and stored in various ways, the two data sources may not be entirely compatible. So a GIS must be able to convert geographic data from one structure to another. In so doing, the implicit assumptions behind different ontologies and classifications require analysis. Object ontologies have gained increasing prominence as a consequence of object-oriented programming and sustained work by Barry Smith and co-workers.

Projections, Coordinate Systems, and Registration

The earth can be represented by various models, each of which may provide a different set of coordinates (e.g., latitude, longitude, elevation) for any given point on the Earth's surface. The simplest model is to assume the earth is a perfect sphere. As more measurements of the earth have accumulated, the models of the earth have become more sophisticated and more accurate. In fact, there are models called datums that apply to different areas of the earth to provide increased accuracy, like NAD83 for U.S. measurements, and the World Geodetic System for worldwide measurements.

Spatial Analysis with Geographical Information System (GIS)

GIS spatial analysis is a rapidly changing field, and GIS packages are increasingly including analytical tools as standard built-in facilities, as optional toolsets, as add-ins or 'analysts'. In many instances these are provided by the original software suppliers (commercial vendors or collaborative non commercial development teams), while in other cases facilities have been developed and are provided by third parties. Furthermore, many products offer software development kits (SDKs), programming languages and language support, scripting facilities and/or special interfaces for developing one's own analytical tools or variants. The website "Geospatial Analysis" and associated book/ebook attempt to provide a reasonably comprehensive guide to the subject. The increased availability has created a new dimension to business intelligence termed "spatial intelligence" which, when openly delivered via intranet, democratizes access to geographic and social network data. Geospatial intelligence, based on GIS spatial analysis, has also become a key element for security. GIS as a whole can be described as conversion to a vectorial representation or to any other digitisation process.

Slope and Aspect

Slope can be defined as the steepness or gradient of a unit of terrain, usually measured as an angle in degrees or as a percentage. Aspect can be defined as the direction in which a unit of terrain faces. Aspect is usually expressed in degrees from north. Slope, aspect, and surface curvature in terrain analysis are all derived from neighborhood operations using elevation values of a cell's adjacent neighbours. Slope is a function of resolution, and the spatial resolution used to calculate slope and aspect should always be specified. Authors such as Skidmore, Jones and Zhou and Liu have compared techniques for calculating slope and aspect.

The following method can be used to derive slope and aspect:

The elevation at a point or unit of terrain will have perpendicular tangents (slope) passing through the point, in an east-west and north-south direction. These two tangents give two components, $\partial z/\partial x$ and $\partial z/\partial y$, which then be used to determine the overall direction of slope, and the aspect of the slope. The gradient is defined as a vector quantity with components equal to the partial derivatives of the surface in the x and y directions.

Data Analysis

It is difficult to relate wetlands maps to rainfall amounts recorded at different points such as airports, television stations, and schools. A GIS, however, can be used to depict two- and three-dimensional characteristics of the Earth's surface, subsurface, and atmosphere from information points. For example, a GIS can quickly generate a map with isopleth or contour lines that indicate differing amounts of rainfall. Such a map can be thought of as a rainfall contour map. Many sophisticated methods can estimate the characteristics of surfaces from a limited number of point measurements. A two-dimensional contour map created from the surface modeling of rainfall point measurements may be overlaid and analyzed with any other map in a GIS covering the same area. This GIS derived map can then provide additional information - such as the viability of water power potential as a renewable energy source. Similarly, GIS can be used to compare other renewable energy resources to find the best geographic potential for a region.

Additionally, from a series of three-dimensional points, or digital elevation model, isopleth lines representing elevation contours can be generated, along with slope analysis, shaded relief, and other elevation products. Watersheds can be easily defined for any given reach, by computing all of the areas contiguous and uphill from any given point of interest. Similarly, an expected thalweg of where surface water would want to travel in intermittent and permanent streams can be computed from elevation data in the GIS.

Topological Modeling

A GIS can recognize and analyze the spatial relationships that exist within digitally stored spatial data. These topological relationships allow complex spatial modelling and analysis to be performed. Topological relationships between geometric entities traditionally include adjacency (what adjoins what), containment (what encloses what), and proximity (how close something is to something else).

Geometric Networks

Geometric networks are linear networks of objects that can be used to represent interconnected features, and to perform special spatial analysis on them. A geometric network is composed of edges, which are connected at junction points, similar to graphs in mathematics and computer science. Just like graphs, networks can have weight and flow assigned to its edges, which can be used to represent various interconnected features more accurately. Geometric networks are often used to model road networks and public utility networks, such as electric, gas, and water networks. Network modeling is also commonly employed in transportation planning, hydrology modeling, and infrastructure modeling.

Hydrological Modeling

GIS hydrological models can provide a spatial element that other hydrological models lack, with the analysis of variables such as slope, aspect and watershed or catchment area. Terrain analysis is fundamental to hydrology, since water always flows down a slope. As basic terrain analysis of a digital elevation model (DEM) involves calculation of slope and aspect, DEMs are very useful for hydrological analysis. Slope and aspect can then be used to determine direction of surface runoff, and hence flow accumulation for the formation of streams, rivers and lakes. Areas of divergent flow can also give a clear indication of the boundaries of a catchment. Once a flow direction and accumulation matrix has been created, queries can be performed that show contributing or dispersal areas at a certain point. More detail can be added to the model, such as terrain roughness, vegetation types and soil types, which can influence infiltration and evapotranspiration rates, and hence influencing surface flow. One of the main uses of hydrological modeling is in environmental contamination research.

Cartographic Modeling

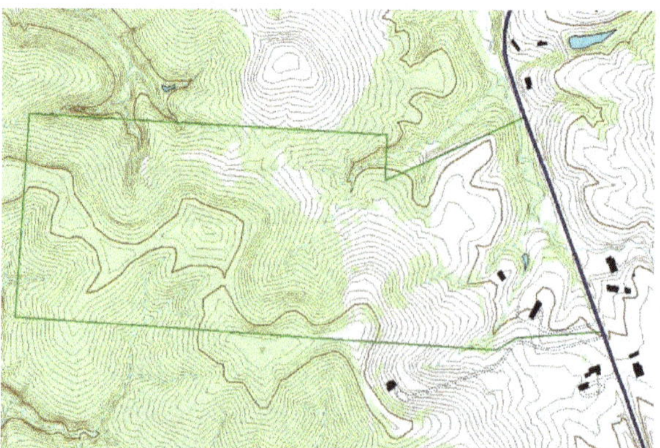

An example of use of layers in a GIS application. In this example, the forest cover layer (light green) is at the bottom, with the topographic layer over it. Next up is the stream layer, then the boundary layer, then the road layer. The order is very important in order to properly display the final result. Note thatthe pond layer was located just below the stream layer, so that a stream line can be seen overlying one of the ponds.

The term "cartographic modeling" was probably coined by Dana Tomlin in his PhD dissertation and later in his book which has the term in the title. Cartographic modeling refers to a process where several thematic layers of the same area are produced, processed, and analyzed. Tomlin used raster layers, but the overlay method can be used more generally. Operations on map layers can be combined into algorithms, and eventually into simulation or optimization models.

Map Overlay

The combination of several spatial datasets (points, lines, or polygons) creates a new output vector dataset, visually similar to stacking several maps of the same region. These overlays are similar to mathematical Venn diagram overlays. A union overlay combines the geographic features and attribute tables of both inputs into a single new output. An intersect overlay defines the area where both inputs overlap and retains a set of attribute fields for each. A symmetric difference overlay defines an output area that includes the total area of both inputs except for the overlapping area.

Data extraction is a GIS process similar to vector overlay, though it can be used in either vector or raster data analysis. Rather than combining the properties and features of both datasets, data extraction involves using a "clip" or "mask" to extract the features of one data set that fall within the spatial extent of another dataset.

In raster data analysis, the overlay of datasets is accomplished through a process known as "local operation on multiple rasters" or "map algebra," through a function that combines the values of each raster's matrix. This function may weigh some inputs more than others through use of an "index model" that reflects the influence of various factors upon a geographic phenomenon.

Geostatistics

Geostatistics is a branch of statistics that deals with field data, spatial data with a continuous index. It provides methods to model spatial correlation, and predict values at arbitrary locations (interpolation).

When phenomena are measured, the observation methods dictate the accuracy of any subsequent analysis. Due to the nature of the data (e.g. traffic patterns in an urban environment; weather patterns over the Pacific Ocean), a constant or dynamic degree of precision is always lost in the measurement. This loss of precision is determined from the scale and distribution of the data collection.

To determine the statistical relevance of the analysis, an average is determined so that points (gradients) outside of any immediate measurement can be included to determine their predicted behavior. This is due to the limitations of the applied statistic and data collection methods, and interpolation is required to predict the behavior of particles, points, and locations that are not directly measurable.

Hillshade model derived from a Digital Elevation Model of the Valestra area in the northern Apennines (Italy)

Interpolation is the process by which a surface is created, usually a raster dataset, through the input of data collected at a number of sample points. There are several forms of interpolation, each which treats the data differently, depending on the properties of the data set. In comparing interpolation methods, the first consideration should be whether or not the source data will change (exact or approximate). Next is whether the method is subjective, a human interpretation, or objective. Then there is the nature of transitions between points: are they abrupt or gradual. Finally, there is whether a method is global (it uses the entire data set to form the model), or local where an algorithm is repeated for a small section of terrain.

Interpolation is a justified measurement because of a spatial autocorrelation principle that recognizes that data collected at any position will have a great similarity to, or influence of those locations within its immediate vicinity.

Digital elevation models, triangulated irregular networks, edge-finding algorithms, Thiessen polygons, Fourier analysis, (weighted) moving averages, inverse distance weighting, kriging, spline, and trend surface analysis are all mathematical methods to produce interpolative data.

Address Geocoding

Geocoding is interpolating spatial locations (X,Y coordinates) from street addresses or any other spatially referenced data such as ZIP Codes, parcel lots and address locations. A reference theme is required to geocode individual addresses, such as a road centerline file with address ranges. The individual address locations have historically been interpolated, or estimated, by examining address ranges along a road segment. These are usually provided in the form of a table or database. The software will then place a dot approximately where that address belongs along the segment of centerline. For example, an address point of 500 will be at the midpoint of a line segment that starts with address 1 and ends with address 1,000. Geocoding can also be applied against actual parcel data, typically from municipal tax maps. In this case, the result of the geocoding will be an actually positioned space as opposed to an interpolated point. This approach is being increasingly used to provide more precise location information.

Reverse Geocoding

Reverse geocoding is the process of returning an estimated street address number as it relates to a given coordinate. For example, a user can click on a road centerline theme (thus providing a coordinate) and have information returned that reflects the estimated house number. This house number is interpolated from a range assigned to that road segment. If the user clicks at the midpoint of a segment that starts with address 1 and ends with 100, the returned value will be somewhere near 50. Note that reverse geocoding does not return actual addresses, only estimates of what should be there based on the predetermined range.

Multi-criteria Decision Analysis

Coupled with GIS, multi-criteria decision analysis methods support decision-makers in analysing a set of alternative spatial solutions, such as the most likely ecological habitat for restoration, against multiple criteria, such as vegetation cover or roads. MCDA uses decision rules to aggregate the criteria, which allows the alternative solutions to be ranked or prioritised. GIS MCDA may reduce costs and time involved in identifying potential restoration sites.

Data Output and Cartography

Cartography is the design and production of maps, or visual representations of spatial data. The vast majority of modern cartography is done with the help of computers, usually using GIS but production of quality cartography is also achieved by importing layers into a design program to refine it. Most GIS software gives the user substantial control over the appearance of the data.

Cartographic work serves two major functions:

First, it produces graphics on the screen or on paper that convey the results of analysis to the people who make decisions about resources. Wall maps and other graphics can be generated, allowing the viewer to visualize and thereby understand the results of analyses or simulations of potential events. Web Map Servers facilitate distribution of generated maps through web browsers using various implementations of web-based application programming interfaces (AJAX, Java, Flash, etc.).

Second, other database information can be generated for further analysis or use. An example would be a list of all addresses within one mile (1.6 km) of a toxic spill.

Graphic Display Techniques

Traditional maps are abstractions of the real world, a sampling of important elements portrayed on a sheet of paper with symbols to represent physical objects. People who use maps must interpret these symbols. Topographic maps show the shape of land surface with contour lines or with shaded relief.

Today, graphic display techniques such as shading based on altitude in a GIS can make relationships among map elements visible, heightening one's ability to extract and analyze information. For example, two types of data were combined in a GIS to produce a perspective view of a portion of San Mateo County, California.

- The digital elevation model, consisting of surface elevations recorded on a 30-meter horizontal grid, shows high elevations as white and low elevation as black.
- The accompanying Landsat Thematic Mapper image shows a false-color infrared image looking down at the same area in 30-meter pixels, or picture elements, for the same coordinate points, pixel by pixel, as the elevation information.

A GIS was used to register and combine the two images to render the three-dimensional perspective view looking down the San Andreas Fault, using the Thematic Mapper image pixels, but shaded using the elevation of the landforms. The GIS display depends on the viewing point of the observer and time of day of the display, to properly render the shadows created by the sun's rays at that latitude, longitude, and time of day.

An archeochrome is a new way of displaying spatial data. It is a thematic on a 3D map that is applied to a specific building or a part of a building. It is suited to the visual display of heat-loss data.

Spatial ETL

Spatial ETL tools provide the data processing functionality of traditional Extract, Transform, Load (ETL) software, but with a primary focus on the ability to manage spatial data. They provide GIS users with the ability to translate data between different standards and proprietary formats, whilst geometrically transforming the data en route. These tools can come in the form of add-ins to existing wider-purpose software such as Microsoft Excel.

GIS Data Mining

GIS or spatial data mining is the application of data mining methods to spatial data. Data mining, which is the partially automated search for hidden patterns in large databases, offers great potential benefits for applied GIS-based decision making. Typical applications include environmental monitoring. A characteristic of such applications is that spatial correlation between data measurements require the use of specialized algorithms for more efficient data analysis.

Applications

The implementation of a GIS is often driven by jurisdictional (such as a city), purpose, or application requirements. Generally, a GIS implementation may be custom-designed for an organization. Hence, a GIS deployment developed for an application, jurisdiction, enterprise, or purpose may not be necessarily interoperable or compatible with a GIS that has been developed for some other application, jurisdiction, enterprise, or purpose.

GIS provides, for every kind of location-based organization, a platform to update geographical data without wasting time to visit the field and update a database manually. GIS when integrated with other powerful enterprise solutions like SAP and the Wolfram Language helps creating powerful decision support system at enterprise level.

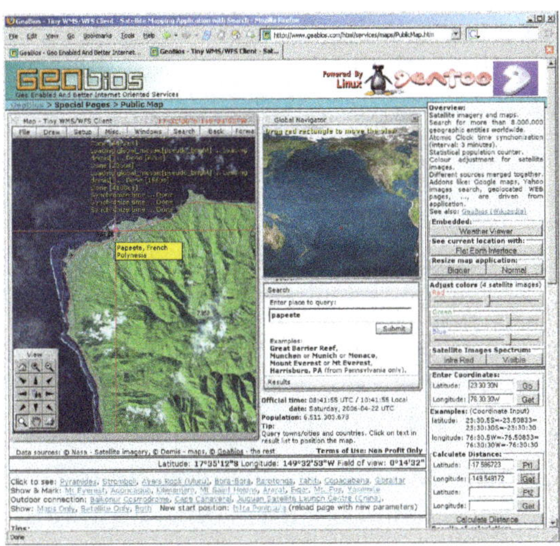

GeaBios – tiny WMS/WFS client (Flash/DHTML)

Many disciplines can benefit from GIS technology. An active GIS market has resulted in lower costs and continual improvements in the hardware and software components of GIS, and usage in the fields of science, government, business, and industry, with applications including real estate, public health, crime mapping, national defense, sustainable development, natural resources, climatology, landscape architecture, archaeology, regional and community planning, transportation and logistics. GIS is also diverging into location-based services, which allows GPS-enabled mobile devices to display their location in relation to fixed objects (nearest restaurant, gas station, fire hydrant) or mobile objects (friends, children, police car), or to relay their position back to a central server for display or other processing.

Open Geospatial Consortium standards

The Open Geospatial Consortium (OGC) is an international industry consortium of 384 companies, government agencies, universities, and individuals participating in a consensus process to develop publicly available geoprocessing specifications. Open interfaces and protocols defined by OpenGIS Specifications support interoperable solutions that "geo-enable" the Web, wireless and location-based services, and mainstream IT, and empower technology developers to make complex spatial information and services accessible and useful with all kinds of applications. Open Geospatial Consortium protocols include Web Map Service, and Web Feature Service.

GIS products are broken down by the OGC into two categories, based on how completely and accurately the software follows the OGC specifications.

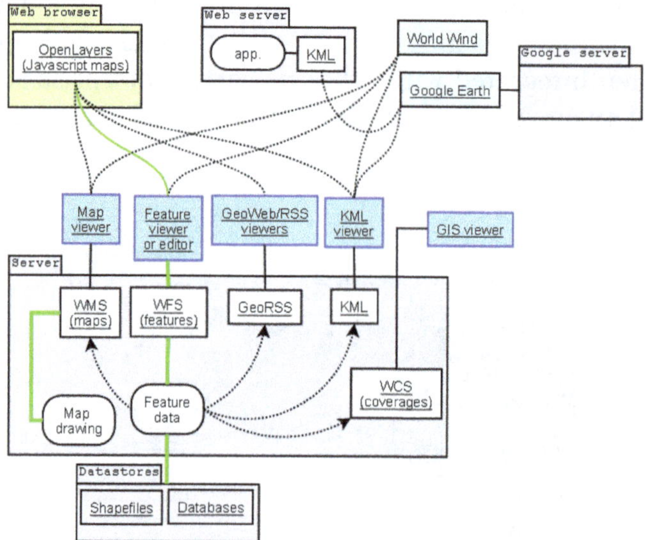

OGC standards help GIS tools communicate.

Compliant Products are software products that comply to OGC's OpenGIS Specifications. When a product has been tested and certified as compliant through the OGC Testing Program, the product is automatically registered as "compliant" on this site.

Implementing Products are software products that implement OpenGIS Specifications but have not yet passed a compliance test. Compliance tests are not available for all specifications. Developers can register their products as implementing draft or approved specifications, though OGC reserves the right to review and verify each entry.

Web Mapping

In recent years there has been an explosion of mapping applications on the web such as Google Maps and Bing Maps. These websites give the public access to huge amounts of geographic data.

Some of them, like Google Maps and OpenLayers, expose an API that enable users to create custom applications. These toolkits commonly offer street maps, aerial/satellite imagery, geocoding, searches, and routing functionality. Web mapping has also uncovered the potential of crowdsourcing geodata in projects like OpenStreetMap, which is a collaborative project to create a free editable map of the world.

Adding the Dimension of Time

The condition of the Earth's surface, atmosphere, and subsurface can be examined by feeding satellite data into a GIS. GIS technology gives researchers the ability to examine the variations in Earth processes over days, months, and years. As an example, the changes in vegetation vigor through a growing season can be animated to determine when drought was most extensive in a particular region. The resulting graphic represents a rough measure of plant health. Working with two variables over time would then allow researchers to detect regional differences in the lag between a decline in rainfall and its effect on vegetation.

GIS technology and the availability of digital data on regional and global scales enable such analyses. The satellite sensor output used to generate a vegetation graphic is produced for example by the Advanced Very High Resolution Radiometer (AVHRR). This sensor system detects the amounts of energy reflected from the Earth's surface across various bands of the spectrum for surface areas of about 1 square kilometer. The satellite sensor produces images of a particular location on the Earth twice a day. AVHRR and more recently the Moderate-Resolution Imaging Spectroradiometer (MODIS) are only two of many sensor systems used for Earth surface analysis. More sensors will follow, generating ever greater amounts of data.

In addition to the integration of time in environmental studies, GIS is also being explored for its ability to track and model the progress of humans throughout their daily routines. A concrete example of progress in this area is the recent release of time-specific population data by the U.S. Census. In this data set, the populations of cities are shown for daytime and evening hours highlighting the pattern of concentration and dispersion generated by North American commuting patterns. The manipulation and generation of data required to produce this data would not have been possible without GIS.

Using models to project the data held by a GIS forward in time have enabled planners to test policy decisions using spatial decision support systems.

Semantics

Tools and technologies emerging from the W3C's Data Activity are proving useful for data integration problems in information systems. Correspondingly, such technologies have been proposed as a means to facilitate interoperability and data reuse among GIS applications. and also to enable new analysis mechanisms.

Ontologies are a key component of this semantic approach as they allow a formal, machine-readable specification of the concepts and relationships in a given domain. This in turn allows a GIS to focus on the intended meaning of data rather than its syntax or structure. For example, reasoning that a land cover type classified as *deciduous needleleaf trees* in one dataset is a specialization or subset of land cover type *forest* in another more roughly classified dataset can help a GIS automatically merge the two datasets under the more general land cover classification. Tentative ontologies have been developed in areas related to GIS applications, for example the hydrology ontology developed by the Ordnance Survey in the United Kingdom and the SWEET ontologies developed by NASA's Jet Propulsion Laboratory. Also, simpler ontologies and semantic metadata standards are being proposed by the W3C Geo Incubator Group to represent geospatial data on the web. GeoSPARQL is a standard developed by the Ordnance Survey, United States Geological Survey, Natural Resources Canada, Australia's Commonwealth Scientific and Industrial Research Organisation and others to support ontology creation and reasoning using well-understood OGC literals (GML, WKT), topological relationships (Simple Features, RCC8, DE-9IM), RDF and the SPARQL database query protocols.

Recent research results in this area can be seen in the International Conference on Geospatial Semantics and the Terra Cognita – Directions to the Geospatial Semantic Web workshop at the International Semantic Web Conference.

Implications of GIS in Society

With the popularization of GIS in decision making, scholars have begun to scrutinize the social and political implications of GIS. GIS can also be misused to distort reality for individual and political gain. It has been argued that the production, distribution, utilization, and representation of geographic information are largely related with the social context and has the potential to increase citizen trust in government. Other related topics include discussion on copyright, privacy, and censorship. A more optimistic social approach to GIS adoption is to use it as a tool for public participation.

Geographic Information Science

Geographic information science or Geographical information science (GIScience) is the scientific discipline that studies data structures and computational techniques to capture, represent, process, and analyze geographic information. It can be contrasted with geographic information systems, which are software tools. British geographer Michael Goodchild has defined this area in the 1990s, and summarized its core interests, including spatial analysis, visualization, and the representation of uncertainty. GIScience is conceptually related to geography, information science, computer science, geomatics and geoinformatics, but it claims the status of an independent scientific discipline.

Definitions

Since its inception in the 1990s, the boundaries between GIScience and cognate disciplines are contested, and different communities might disagree on what GIScience is and what it studies. In particular, Goodchild stated that "information science can be defined as the systematic study according to scientific principles of the nature and properties of information. Geographic information science is the subset of information science that is about geographic information." Another influential definition is that by GIScientist David Mark, which states:

Geographic Information Science (GIScience) is the basic research field that seeks to redefine geographic concepts and their use in the context of geographic information systems. GIScience also examines the impacts of GIS on individuals and society, and the influences of society on GIS. GIScience re-examines some of the most fundamental themes in traditional spatially oriented fields such as geography, cartography, and geodesy, while incorporating more recent developments in cognitive and information science. It also overlaps with and draws from more specialized research fields such as computer science, statistics, mathematics, and psychology, and contributes to progress in those fields. It supports research in political science and anthropology, and draws on those fields in studies of geographic information and society.

Spatial Data Infrastructure

A spatial data infrastructure (SDI) is a data infrastructure implementing a framework of geographic data, metadata, users and tools that are interactively connected in order to use spatial data in an efficient and flexible way. Another definition is "the technology, policies, standards, human resources, and related activities necessary to acquire, process, distribute, use, maintain, and preserve spatial data".

A further definition is given in Kuhn (2005): "An SDI is a coordinated series of agreements on technology standards, institutional arrangements, and policies that enable the discovery and use of geospatial information by users and for purposes other than those it was created for."

General

Some of the main principles are that data and metadata should not be managed centrally, but by the data originator and/or owner, and that tools and services connect via computer networks to the various sources. A GIS is often the platform for deploying an individual node within an SDI. To achieve these objectives, good coordination between all the actors is necessary and the definition of standards is very important.

Due to its nature (size, cost, number of interactors) an SDI is usually government-related. An example of an existing SDI is the National Spatial Data Infrastructure (NSDI) in the United States. At the European side, INSPIRE is a European Commission initiative to build a European SDI beyond national boundaries and ultimately the United Nations Spatial Data Infrastructure (UNSDI) will do the same for over 30 UN Funds, Programmes, Specialized Agencies and member countries.

Software Components

A SDI should enable the discovery and delivery of spatial data from a data repository, via a spatial service provider, to a user. As mentioned earlier it is often wished that the data provider is able to update spatial data stored in a repository. Hence, the basic software components of an SDI are:

- Software client - to display, query, and analyse spatial data (this could be a browser or a desktop GIS)
- Catalogue service - for the discovery, browsing, and querying of metadata or spatial services, spatial datasets and other resources
- Spatial data service - allowing the delivery of the data via the Internet
- Processing services - such as datum and projection transformations
- (Spatial) data repository - to store data, e.g., a spatial database
- GIS software (client or desktop) - to create and update spatial data

Besides these software components, a range of (international) technical standards are necessary that allow interaction between the different software components. Among those are geospatial standards defined by the Open Geospatial Consortium (e.g., OGC WMS, WFS, GML etc.) and ISO (e.g., ISO 19115) for the delivery of maps, vector and raster data, but also data format and internet transfer standards by W3C consortium.

Spatial Database

A spatial database is a database that is optimized to store and query data that represents objects defined in a geometric space. Most spatial databases allow representing simple geometric objects such as points, lines and polygons. Some spatial databases handle more complex structures such as 3D objects, topological coverages, linear networks, and TINs. While typical databases have developed to manage various numeric and character types of data, such databases require additional functionality to process spatial data types efficiently, and developers have often added *geometry* or *feature* data types. The Open Geospatial Consortium developed the Simple Features specification (first released in 1997) and sets standards for adding spatial functionality to database systems. The *SQL/MM Spatial* ISO/EIC standard is a part the SQL/MM multimedia standard and extends the Simple Features standard with data types that support circular interpolations.

Geodatabase

A geodatabase (also geographical database and geospatial database) is a database of geographic data, such as countries, administrative divisions, cities, and related information. Such databases can be useful for websites that wish to identify the locations of their visitors for customization purposes.

Features of Spatial Databases

Database systems use indexes to quickly look up values and the way that most databases index data is not optimal for spatial queries. Instead, spatial databases use a spatial index to speed up database operations.

In addition to typical SQL queries such as SELECT statements, spatial databases can perform a wide variety of spatial operations. The following operations and many more are specified by the Open Geospatial Consortium standard:

- Spatial Measurements: Computes line length, polygon area, the distance between geometries, etc.
- Spatial Functions: Modify existing features to create new ones, for example by providing a buffer around them, intersecting features, etc.
- Spatial Predicates: Allows true/false queries about spatial relationships between geometries. Examples include "do two polygons overlap" or 'is there a residence located within a mile of the area we are planning to build the landfill?'
- Geometry Constructors: Creates new geometries, usually by specifying the vertices (points or nodes) which define the shape.
- Observer Functions: Queries which return specific information about a feature such as the location of the center of a circle

Some databases support only simplified or modified sets of these operations, especially in cases of NoSQL systems like MongoDB and CouchDB.

Spatial Index

Spatial indices are used by spatial databases (databases which store information related to objects in space) to optimize spatial queries. Conventional index types do not efficiently handle spatial queries such as how far two points differ, or whether points fall within a spatial area of interest. Common spatial index methods include:

- Grid (spatial index)
- Z-order (curve)
- Quadtree
- Octree
- UB-tree
- R-tree: Typically the preferred method for indexing spatial data.Objects (shapes, lines and points) are grouped using the minimum bounding rectangle (MBR). Objects are added to an MBR within the index that will lead to the smallest increase in its size.
- R+ tree
- R* tree
- Hilbert R-tree
- X-tree
- kd-tree
- m-tree – an m-tree index can be used for the efficient resolution of similarity queries on complex objects as compared using an arbitrary metric.
- Point access method
- Binary space partitioning (BSP-Tree): Subdividing space by hyperplanes.

Spatial Database Systems

List

- All OpenGIS specifications compliant products
- Open-source spatial databases and APIs, some of which are OpenGIS-compliant
- Caliper extends the Raima Data Manager with spatial datatypes, functions, and utilities.
- Boeing's Spatial Query Server spatially enables Sybase ASE.
- Smallworld VMDS, the native GE Smallworld GIS database

- SpatiaLite extends Sqlite with spatial datatypes, functions, and utilities.
- IBM DB2 Spatial Extender can spatially-enable any edition of DB2, including the free DB2 Express-C, with support for spatial types
- ClusterPoint offers native indexed support for distances, range matching and polygon matching, as well as aggregation.
- Oracle Spatial
- Oracle Locator
- Vertica Place, the geo-spatial extension for HP Vertica, adds OGC-compliant spatial features to the relational column-store database.
- Microsoft SQL Server has support for spatial types since version 2008
- PostgreSQL DBMS (database management system) uses the spatial extension PostGIS to implement the standardized datatype *geometry* and corresponding functions.
- Teradata Geospatial includes 2D spatial functionality (OGC-compliant) in its data warehouse system.
- MonetDB/GIS extension for MonetDB adds OGS Simple Features to the relational column-store database.
- Linter SQL Server supports spatial types and spatial functions according to the OpenGIS specifications.
- MySQL DBMS implements the datatype *geometry*, plus some spatial functions implemented according to the OpenGIS specifications. However, in MySQL version 5.5 and earlier, functions that test spatial relationships are limited to working with minimum bounding rectangles rather than the actual geometries. MySQL versions earlier than 5.0.16 only supported spatial data in MyISAM tables. As of MySQL 5.0.16, InnoDB, NDB, BDB, and ARCHIVE also support spatial features.
- Neo4j – a graph database that can build 1D and 2D indexes as B-tree, Quadtree and Hilbert curve directly in the graph
- AllegroGraph – a graph database which provides a novel mechanismfor efficient storage and retrieval of two-dimensional geospatial coordinates for Resource Description Framework data. It includes an extension syntax for SPARQL queries.
- MarkLogic, MongoDB, RavenDB, and RethinkDB support geospatial indexes in 2D.
- Esri has a number of both single-user and multiuser geodatabases.

- SpaceBase, a real-time spatial database.
- CouchDB a document-based database system that can be spatially enabled by a plugin called Geocouch
- CartoDB, a cloud-based geospatial database on top of PostgreSQL with PostGIS
- StormDB, an upcoming cloud-based database on top of PostgreSQL with geospatial capabilities
- AsterixDB, an open-source big data management system with native geospatial capabilities
- SpatialDB by MineRP, the world's first open-standards (OGC) spatial database with spatial type extensions for the Mining Industry
- H2 supports geometry types and spatial indices as of version 1.3.173 (2013-07-28). An extension called H2GIS available on Maven Central gives full OGC Simple Features support.
- GeoMesa is a cloud-based spatio-temporal database built on top of Apache Accumulo and Apache Hadoop. GeoMesa supports full OGC Simple Features support and a GeoServer plugin.
- Ingres 10S and 10.2 include native comprehensive spatial support. Ingres includes the Geospatial Data Abstraction Library cross-platform spatial data translator.
- Tarantool supports geospatial queries with RTREE index.
- SAP HANA supports geospatial with SPS08 .
- Redis with the Geo API

References

- Monmonier, Mark (1991). How to Lie with Maps. Chicago, IL: University of Chicago Press. ISBN 0226534219.
- Duckham, Matt; Goodchild, Michael F.; Worboys, Michael (2004-11-23). Foundations of Geographic Information Science. CRC Press. ISBN 9780203009543.
- Duckham, Matt; Goodchild, Michael F.; Worboys, Michael (2004-11-23). Foundations of Geographic Information Science. CRC Press. p. 4. ISBN 9780203009543.
- (eds.), Wolfgang Kresse, David M. Danko (2010). Springer handbook of geographic information (1. Ed. ed.). Berlin: Springer. pp. 82–83. ISBN 9783540726807.
- "Integrating GIS with SAP--The Imperative - ArcNews Spring 2009 Issue". www.esri.com. Retrieved 2016-07-13.
- Monmonier, Mark (2005). "Lying with Maps". Statistical Science. Retrieved 2016. Check date

values in: |access-date= (help)

- The Remarkable History of GIS - Geographical Information Systems."The Remarkable History of GIS". Retrieved 2015-05-05.
- Winther, Rasmus G. (2014) "Mapping Kinds in GIS and Cartography" in Natural Kinds and Classification in Scientific Practice, edited by C. Kendig http://philpapers.org/archive/WINMKI.pdf
- "Aeryon Announces Version 5 of the Aeryon Scout System | Aeryon Labs Inc". Aeryon.com. 2011-07-06. Retrieved 2012-05-13.
- "Geospatial Analysis – a comprehensive guide. 2nd edition © 2006–2008 de Smith, Goodchild, Longley". Spatialanalysisonline.com. Retrieved 2012-05-13.
- "Rapport sur la marche et les effets du choléra dans Paris et le département de la Seine. Année 1832". Gallica. Retrieved 10 May 2012.
- Greene, R.; Devillers, R.; Luther, J.E.; Eddy, B.G. (2011). "GIS-based multi-criteria analysis". Geography Compass. 5/6: 412–432.
- Maliene V, Grigonis V, Palevičius V, Griffiths S (2011). "Geographic information system: Old principles with new capabilities". Urban Design International. pp. 1–6. doi:10.1057/udi.2010.25.

2

Essential Elements of Geographic Information Systems

The major elements of geographic information system are GIS file formats, conservation geoportal, data model, digital mapping, distributed GIS etc. A GIS file format is the conversion of geographical information into a computer file. Geographical information systems need a complete listing of data on an online portal. The online portal for these data sets is known as conservation geoportal. The topics discussed in the chapter are of great importance to broaden the existing knowledge on geographic information system.

GIS File Formats

A GIS file format is a standard of encoding geographical information into a computer file. They are created mainly by government mapping agencies (such as the USGS or National Geospatial-Intelligence Agency) or by GIS software developers.

Raster

A raster data type is, in essence, any type of digital image represented by reducible and enlargeable grids. Anyone who is familiar with digital photography will recognize the Raster graphics pixel as the smallest individual grid unit building block of an image, usually not readily identified as an artifact shape until an image is produced on a very large scale. A combination of the pixels making up an image color formation scheme will compose details of an image, as is distinct from the commonly used points, lines, and polygon area location symbols of scalable vector graphics as the basis of the vector model of area attribute rendering. While a digital image is concerned with its output blending together its grid based details as an identifiable representation of reality, in a photograph or art image transferred into a computer, the raster data type will reflect a digitized abstraction of reality dealt with by grid populating tones or objects, quantities, cojoined or open boundaries, and map relief schemas. Aerial photos are one commonly used form of raster data, with one primary purpose in mind: to display a detailed image on a map area, or for the purposes of rendering its identifiable objects by digitization. Additional raster data sets used by a GIS will contain information regarding elevation, a digital elevation model, or reflectance of a particular wavelength of light, Landsat, or other electromagnetic spectrum indicators.

Raster data type consists of rows and columns of cells, with each cell storing a single value. Raster data can be images raster images) with each pixel (or cell) containing a color value. Additional values recorded for each cell may be a discrete value, such as land use, a continuous value, such as temperature, or a null value if no data is available. While a raster cell stores a single value, it can be extended by using raster bands to represent RGB (red, green, blue) colors, colormaps (a mapping between a thematic code and RGB value), or an extended attribute table with one row for each unique cell value. The resolution of the raster data set is its cell width in ground units.

Digital elevation model, map (image), and vector data

Raster data is stored in various formats; from a standard file-based structure of TIFF, JPEG, etc. to binary large object (BLOB) data stored directly in a relational database management system (RDBMS) similar to other vector-based feature classes. Database storage, when properly indexed, typically allows for quicker retrieval of the raster data but can require storage of millions of significantly sized records.

Vector

In a GIS, geographical features are often expressed as vectors, by considering those features as geometrical shapes. Different geographical features are expressed by different types of geometry:

- Points

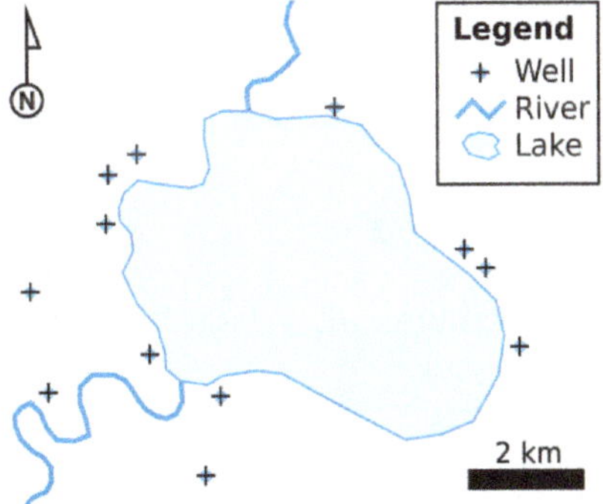

A simple vector map, using each of the vector elements: points for wells, lines for rivers, and a polygon for the lake

Zero-dimensional points are used for geographical features that can best be expressed by a single point reference—in other words, by simple location. Examples include wells, peaks, features of interest, and trailheads. Points convey the least amount of information of these file types. Points can also be used to represent areas when displayed at a small scale. For example, cities on a map of the world might be represented by points rather than polygons. No measurements are possible with point features.

- Lines or polylines

  One-dimensional lines or polylines are used for linear features such as rivers, roads, railroads, trails, and topographic lines. Again, as with point features, linear features displayed at a small scale will be represented as linear features rather than as a polygon. Line features can measure distance.

- Polygons

  Two-dimensional polygons are used for geographical features that cover a particular area of the earth's surface. Such features may include lakes, park boundaries, buildings, city boundaries, or land uses. Polygons convey the most amount of information of the file types. Polygon features can measure perimeter and area.

Each of these geometries are linked to a row in a database that describes their attributes. For example, a database that describes lakes may contain a lake's depth, water quality, pollution level. This information can be used to make a map to describe a particular attribute of the dataset. For example, lakes could be coloured depending on level of pollution. Different geometries can also be compared. For example, the GIS could be used to identify all wells (point geometry) that are within one kilometre of a lake (polygon geometry) that has a high level of pollution.

Vector features can be made to respect spatial integrity through the application of topology rules such as 'polygons must not overlap'. Vector data can also be used to represent continuously varying phenomena. Contour lines and triangulated irregular networks (TIN) are used to represent elevation or other continuously changing values. TINs record values at point locations, which are connected by lines to form an irregular mesh of triangles. The face of the triangles represent the terrain surface.

Advantages and Disadvantages

There are some important advantages and disadvantages to using a raster or vector data model to represent reality:

- Raster datasets record a value for all points in the area covered which may require more storage space than representing data in a vector format that can store data only where needed.

- Raster data is computationally less expensive to render than vector graphics
- There are transparency and aliasing problems when overlaying multiple stacked pieces of raster images
- Vector data allows for visually smooth and easy implementation of overlay operations, especially in terms of graphics and shape-driven information like maps, routes and custom fonts, which are more difficult with raster data.
- Vector data can be displayed as vector graphics used on traditional maps, whereas raster data will appear as an image that may have a blocky appearance for object boundaries. (depending on the resolution of the raster file)
- Vector data can be easier to register, scale, and re-project, which can simplify combining vector layers from different sources.
- Vector data is more compatible with relational database environments, where they can be part of a relational table as a normal column and processed using a multitude of operators.
- Vector file sizes are usually smaller than raster data, which can be tens, hundreds or more times larger than vector data (depending on resolution).
- Vector data is simpler to update and maintain, whereas a raster image will have to be completely reproduced. (Example: a new road is added).
- Vector data allows much more analysis capability, especially for "networks" such as roads, power, rail, telecommunications, etc. (Examples: Best route, largest port, airfields connected to two-lane highways). Raster data will not have all the characteristics of the features it displays.

Non-Spatial Data

Additional non-spatial data can also be stored along with the spatial data represented by the coordinates of a vector geometry or the position of a raster cell. In vector data, the additional data contains attributes of the feature. For example, a forest inventory polygon may also have an identifier value and information about tree species. In raster data the cell value can store attribute information, but it can also be used as an identifier that can relate to records in another table.

Software is currently being developed to support spatial and non-spatial decision-making, with the solutions to spatial problems being integrated with solutions to non-spatial problems. The end result with these flexible spatial decision-making support systems (FSDSSs) is expected to be that non-experts will be able to use GIS, along with spatial criteria, and simply integrate their non-spatial criteria to view solutions to multi-criteria problems. This system is intended to assist decision-making.

Popular GIS File Formats

Raster Formats

- ADRG – National Geospatial-Intelligence Agency (NGA)'s ARC Digitized Raster Graphics
- Binary file – An unformatted file consisting of raster data written in one of several data types, where multiple band are stored in BSQ (band sequential), BIP (band interleaved by pixel) or BIL (band interleaved by line). Georeferencing and other metadata are stored one or more sidecar files.
- Digital raster graphic (DRG) – digital scan of a paper USGS topographic map
- ECRG – National Geospatial-Intelligence Agency (NGA)'s Enhanced Compressed ARC Raster Graphics (Better resolution than CADRG and no color loss)
- ECW – Enhanced Compressed Wavelet (from ERDAS). A compressed wavelet format, often lossy.
- Esri grid – proprietary binary and metadataless ASCII raster formats used by Esri
- GeoTIFF – TIFF variant enriched with GIS relevant metadata
- IMG – ERDAS IMAGINE image file format
- JPEG2000 – Open-source raster format. A compressed format, allows both lossy and lossless compression.
- MrSID – Multi-Resolution Seamless Image Database (by Lizardtech). A compressed wavelet format, allows both lossy and lossless compression.
- netCDF-CF – netCDF file format with CF medata conventions for earth science data. Binary storage in open format with optional compression. Allows for direct web-access of subsets/aggregations of maps through OPeNDAP protocol.
- RPF – Raster Product Format, military file format specified in MIL-STD-2411
    - CADRG – Compressed ADRG, developed by NGA, nominal compression of 55:1 over ADRG (type of Raster Product Format)
    - CIB – Controlled Image Base, developed by NGA (type of Raster Product Format)

Vector Formats

- AutoCAD DXF – contour elevation plots in AutoCAD DXF format (by Autodesk)
- Cartesian coordinate system (XYZ) – simple point cloud

- Digital line graph (DLG) – a USGS format for vector data
- Esri TIN - proprietary binary format for triangulated irregular network data used by Esri
- Geography Markup Language (GML) – XML based open standard (by OpenGIS) for GIS data exchange
- GeoJSON – a lightweight format based on JSON, used by many open source GIS packages
- GeoMedia – Intergraph's Microsoft Access based format for spatial vector storage
- ISFC – Intergraph's MicroStation based CAD solution attaching vector elements to a relational Microsoft Access database
- Keyhole Markup Language (KML) – XML based open standard (by OpenGIS) for GIS data exchange
- MapInfo TAB format – MapInfo's vector data format using TAB, DAT, ID and MAP files
- National Transfer Format (NTF) – National Transfer Format (mostly used by the UK Ordnance Survey)
- Spatialite – is a spatial extension to SQLite, providing vector geodatabase functionality. It is similar to PostGIS, Oracle Spatial, and SQL Server with spatial extensions
- Shapefile – a popular vector data GIS format, developed by Esri
- Simple Features – Open Geospatial Consortium specification for vector data
- SOSI – a spatial data format used for all public exchange of spatial data in Norway
- Spatial Data File – Autodesk's high-performance geodatabase format, native to MapGuide
- TIGER – Topologically Integrated Geographic Encoding and Referencing
- Vector Product Format (VPF) – National Geospatial-Intelligence Agency (NGA)'s format of vectored data for large geographic databases

Grid Formats (For Elevation)

- USGS DEM – The USGS' Digital Elevation Model
    - GTOPO30 – Large complete Earth elevation model at 30 arc seconds, delivered in the USGS DEM format

- DTED – National Geospatial-Intelligence Agency (NGA)'s Digital Terrain Elevation Data, the military standard for elevation data
- GeoTIFF – TIFF variant enriched with GIS relevant metadata
- SDTS – The USGS' successor to DEM

Other Formats

- Dual Independent Map Encoding (DIME) – A historic GIS file format, developed in the 1960s
- Geographic Data Files (GDF) — An interchange file format for geographic data
- GeoPackage (GPKG) – An standards-based open format based on the SQLite database format for both vector and raster data
- Well-known text (WKT) – A text markup language for representing feature geometry, developed by Open Geospatial Consortium
- Well-known binary (WKB) – Binary version of Well-known text
- World file – Georeferencing a raster image file (e.g. JPEG, BMP)

Conservation Geoportal

The Conservation Geoportal was an online geoportal, intended to provide a comprehensive listing of geographic information systems (GIS) datasets and web map service relevant to biodiversity conservation. It is currently defunct. The site, its contents and functionality were free for anyone to use and contribute to. The Conservation Geoportal was launched on June 28, 2006 at the joint Society for Conservation Biology and Society for Conservation GIS Conference in San Jose, California, USA. As of October 2007, it included metadata for over 3,667 GIS records.

History

The Conservation Geoportal was conceived when representatives from a group of conservation-minded organizations met at the National Geographic Society in March 2005 to define a vision for a World Conservation Base Map. Initially the focus on developing an inventory or catalog of datasets and maps in the form of a metadata database was to be mined to develop the Conservation Base Map and Atlas.

Overview

The Conservation Geoportal constitutes a collaborative effort by and for the conservation community to facilitate the discovery and publishing of GIS data and maps, to support conservation decision-making and education. It does not actually store maps and data, but rather the descriptions and links to those data resources. These descriptions are known as metadata. It was intended to provide an efficient point of access for people interested in a full range of conservation-related GIS data. Capabilities of the Conservation Geoportal included:

- Search for data and maps by keyword, category, geography, or time period
- Save search queries for future use
- Use the built-in Map Viewer to display, manipulate, and combine live map services
- Map viewer supports OpenGIS standards (WMS, WFS, WCS) and ArcIMS services
- Create, save, and email custom maps using data from various web map service
- Publish metadata for maps and data so others can find them
- Featured Map section
- Content in designated thematic data channels
- Share information with other geoportal

Status

- Sponsored by The Nature Conservancy, National Geographic Society and UNEP-World Conservation Monitoring Centre
- ~2,000 visitors per month at its peak
- ~3,667 metadata records & 515 registered users

Data Channels

The Conservation Geoportal included Data Channels and Sub-channels to organize and facilitate access to metadata describing data and maps in a given topic or theme. Channels provided quick access (2 clicks to content) to key data resources that experts consider important to the larger user community. Channels were managed by organizations and experts (channel stewards) knowledgeable about that theme, including:

- Conservation areas: Conservation areas can include existing legally protected areas, as well as areas of ecological or cultural significance identified through assessment and planning efforts. They represent areas where conservation activities are currently taking place or where one or more organizations intend to take action

- Species: Species distributions including amphibian, birds, fish, mammals and many others
- Habitats: Habitats and ecosystems
- Threats: Threats to biodiversity
- Environmental factors: Physical environmental factors including soils, geology, land cover/land use and oceanography
- Socioeconomics: Factors including population, economy, policy, culture, indigenous rights, ecosystem services
- Base map layers: Layers including roads, political boundaries, and satellite imagery

Geoportal Consortium

The Conservation Geoportal was designed and maintained collaboratively by a consortium of institutions including (in alphabetical order):

- American Museum of Natural History
- Conservation International
- Environmental Systems Research Institute
- IUCN - The World Conservation Union
- NASA
- National Geographic Society
- NatureServe
- Smithsonian Institution
- The Nature Conservancy
- UNEP - World Conservation Monitoring Centre
- University of Maryland - Global Land Cover Facility
- USGS - National Biological Information Infrastructure
- Waterborne Environmental, Inc.
- Wildlife Conservation Society

- World Resources Institute
- World Wildlife Fund

Technology

The Conservation Geoportal was based on ESRI's GIS Portal Toolkit (Version 3.0) and ArcWeb Services technologies. Currently the site is maintained and hosted by ESRI. Although the underlying technology was proprietary, the Conservation Geoportal supports several metadata standards and OpenGIS standards, as:

- FGDC and ISO 19115 metadata standards
- Harvesting from ArcIMS, Z39.50, OAI, and WAF based metadata repositories
- OpenGIS WFS, WMS, and WCS services through the map viewer
- ArcIMS Image and ArcGIS Image Server
- OpenLS geocoder

Mashup Capabilities

The Map Viewer let users overlay or mashup data layers from different map servers, which may be hosted by different organizations using different protocols (e.g., ArcIMS, WMS). For example, by searching the catalog, users could discover three different map services, hosted by the Nature Conservancy, Conservation International, and World Wildlife Fund, delineating conservation priority areas. Then, with a click, these live maps can be overlaid together in the Map Viewer along with a satellite image backdrop from NASA. Users could then zoom and pan to their area of interest, turn layers on and off, adjust transparencies, and save that map view to a URL, which they can e-mail to their colleagues to show how various priority maps compare. When their colleagues click the link, exactly the same map view opens, allowing them to work with the live map, perhaps adding map services or posting the link to the map view on their Web site.

Parent Project

The Conservation Geoportal was intended to support the principles and objectives of the Conservation Commons. At its simplest, it encourages organizations and individuals alike to ensure open access to data, information, expertise and knowledge related to the conservation of biodiversity. The Conservation Commons is the expression of a cooperative effort of non-governmental organizations, international and multi-lateral organizations, governments, academia, and the private sector, to improve open access to and unrestricted use of, data, information and knowledge related to the conservation of biodiversity with the belief that this will contribute to improving conservation outcomes.

Data Model (GIS)

A data model in geographic information systems is a mathematical construct for representing geographic objects or surfaces as data. For example, the vector data model represents geography as collections of points, lines, and polygons; the raster data model represent geography as cell matrices that store numeric values; and the TIN data model represents geography as sets of contiguous, nonoverlapping triangles.

Representing Three-dimensional Map Information

There are two approaches for representing three-dimensional map information, and for managing it in the data model.

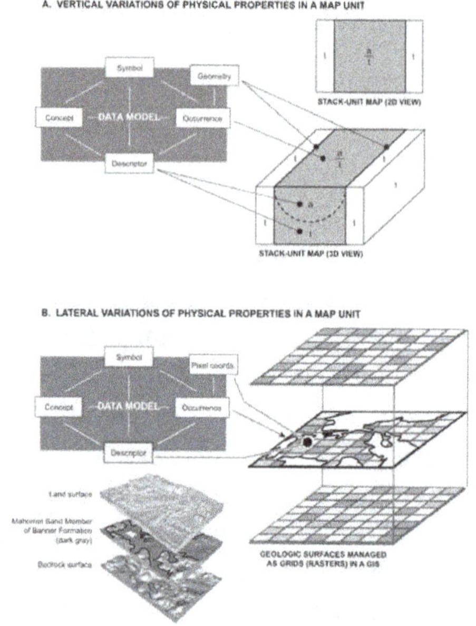

Approaches for representing three-dimensional map information, and for managing it in the data model.

Vector-based Stack-Unit

Vector-based stack-unit maps depict the vertical succession of geologic units to a specified depth (here, the base of the block diagram). This mapping approach characterizes the vertical variations of physical properties in each 3-D map unit. In this example, an alluvial deposit (unit "a") overlies glacial till (unit "t"), and the stack-unit labeled "a/t" indicates that relationship, whereas the unit "t" indicates that glacial till extends down to the specified depth. In a manner similar to that shown in figure 11, the stack-unit's occurrence (the map unit's outcrop), geometry (the map unit's boundaries), and descriptors (the physical properties of the geologic units included in the stack-unit) are managed as they are for a typical 2-D geologic map.

Raster-based Stacked Surfaces

Raster-based stacked surfaces depict the surface of each buried geologic unit, and can accommodate data on lateral variations of physical properties. In this example from Soller and others (1999), the upper surface of each buried geologic unit was represented in raster format as an ArcInfo Grid file. The middle grid is the uppermost surface of an economically important aquifer, the Mahomet Sand, which fills a pre- and inter-glacial valley carved into the bedrock surface. Each geologic unit in raster format can be managed in the data model, in a manner not dissimilar from that shown for the stack-unit map. The Mahomet Sand is continuous in this area, and represents one occurrence of this unit in the data model. Each raster, or pixel, on the Mahomet Sand surface has a set of map coordinates that are recorded in a GIS (in the data model bin that is labeled "Pixel coordinates", which is the raster corollary of the "Geometry" bin for vector map data). Each pixel can have a unique set of descriptive information, such as surface elevation, unit thickness, lithology, transmissivity, etc.).

Digital Mapping

Digital mapping (also called digital cartography) is the process by which a collection of data is compiled and formatted into a virtual image. The primary function of this technology is to produce maps that give accurate representations of a particular area, detailing major road arteries and other points of interest. The technology also allows the calculation of distances from one place to another.

Though digital mapping can be found in a variety of computer applications, such as Google Earth, the main use of these maps is with the Global Positioning System, or GPS satellite network, used in standard automotive navigation systems.

History

From Paper to Paperless

The roots of digital mapping lie within traditional paper maps such as the Thomas Guide. Paper maps provide basic landscapes similar to digitized road maps, yet are often cumbersome, cover only a designated area, and lack many specific details such as road blocks. In addition, there is no way to "update" a paper map except to obtain a new version. On the other hand, digital maps, in many cases, can be updated through synchronization with updates from company servers.

Expanded Capabilities

Early digital maps had the same basic functionality as paper maps—that is, they provided a "virtual view" of roads generally outlined by the terrain encompassing the surrounding area. However, as digital maps have grown with the expansion of GPS technology in the past decade, live traffic updates, points of interest and service locations have been added to enhance digital maps to be more "user conscious." Traditional "virtual views" are now only part of digital mapping. In many cases, users can choose between virtual maps, satellite (aerial views), and hybrid (a combination of virtual map and aerial views) views. With the ability to update and expand digital mapping devices, newly constructed roads and places can be added to appear on maps.

Data Collection

Digital maps heavily rely upon a vast amount of data collected over time. Most of the information that comprise digital maps is the culmination of satellite imagery as well as street level information. Maps must be updated frequently to provide users with the most accurate reflection of a location. While there is a wide spectrum on companies that specialize in digital mapping, the basic premise is that digital maps will accurately portray roads as they actually appear to give "life-like experiences."

Functionality and use

Computer Applications

Computer programs and applications such as Google Earth and Google Maps provide map views from space and street level of much of the world. Used primarily for recreational use, Google Earth provides digital mapping in personal applications, such as tracking distances or finding locations.

Scientific Applications

The development of mobile computing (PDAs, tablet PCs, laptops, etc.) has recently (since about 2000) spurred the use of digital mapping in the sciences and applied sciences. As of 2009, science fields that use digital mapping technology include geology, engineering, architecture, land surveying, mining, forestry, environmental, and archaeology.

GPS Navigation Systems

The principal use by which digital mapping has grown in the past decade has been its connection to Global Positioning System (GPS) technology. GPS is the foundation behind digital mapping navigation systems.

How it Works

The coordinates and position as well as atomic time obtained by a terrestrial GPS receiver from GPS satellites orbiting Earth interact together to provide the digital mapping programming with points of origin in addition to the destination points needed to calculate distance. This information is then analyzed and compiled to create a map that provides the easiest and most efficient way to reach a destination.

More technically speaking, the device operates in the following manner:

1. GPS receivers collect data from at least four GPS satellites orbiting the Earth, calculating position in three dimensions.
2. The GPS receiver then utilizes position to provide GPS coordinates, or exact points of latitudinal and longitudinal direction from GPS satellites.
3. The points, or coordinates, output an accurate range between approximately "10-20 meters" of the actual location.
4. The beginning point, entered via GPS coordinates, and the ending point, (address or coordinates) input by the user, are then entered into the digital mapping software.
5. The mapping software outputs a real-time visual representation of the route. The map then moves along the path of the driver.
6. If the driver drifts from the designated route, the navigation system will use the current coordinates to recalculate a route to the destination location.

Distributed GIS

Distributed GIS concerns itself with GI Systems that do not have all of the system components in the same physical location. This could be the processing, the database, the rendering or the user interface. Examples of distributed systems are web-based GIS, Mobile GIS, Corporate GIS and GRID computing.

Etymology

The term Distributed GIS was coined by Bruce Gittings at the University of Edinburgh. He was responsible for one of the first Internet-based distributed GIS. 1n 1994, he designed and implemented the World Wide Earthquake Locator, which provided maps of recent earthquake occurrences to a location-independent user, which used the Xerox PARC mapping system (based in California, USA), managed by an interface based in Edinburgh (Scotland), which drew data in real-time from the National Earthquake Information Center (USGS) in Colorado, USA. Gittings first taught a course in this subject in 2005 as part of the Masters Programme in GIS at that institution. There being no Wikipedia article relating to Distributed GIS, he set his students the task of creating one in 2007 as a class exercise.

Corporate GIS

A corporate Geographical Information System, is similar to Enterprise GIS and satisfies the spatial information needs of an organisation as a whole in an integrated manner (Chan & Williamson 1997). Corporate GIS consists of four technological elements which are data, standards, information technology and personnel with expertise. It is a coordinated approach that moves away from fragmented desktop GIS. The design of a corporate GIS includes the construction of a centralised corporate database that is designed to be the principle resource for an entire organisation. The corporate database is specifically designed to efficiently and effectively suit the requirements of the organisation. Essential to a corporate GIS is the effective management of the corporate database and the establishment of standards such as OGC for mapping and database technologies.

Benefits

There are many advantages of a corporate GIS. Firstly, all the users in the organisation have access to shared, complete, accurate, high quality and up-to-date data. All the users in the organisation also have access to shared technology and people with expertise. Consequently, this improves the efficiency and effectiveness of the organisation as a whole. A successfully managed corporate database reduces redundant collection and storage of information across the organisation. By centralising resources and efforts, it reduces the overall cost.

Recommended Use

A corporate GIS is recommended for anyone from local governments to global governmental organisations. This is particularly useful if data is to be shared between governmental departments or organisations. However, a corporate GIS is considered not to be cost efficient for smaller organisations as it is expensive to implement.

Mobile GIS

The number of mobile devices in circulation has surpassed the world's population (2013) with a rapid acceleration in iOS, Android and Windows 8 tablet up-take. Tablets are fast becoming popular for Utility field use. Low-cost MIL-STD-810 certified cases transform consumer tablets into fully ruggedised, yet lightweight field use units at 10% of legacy ruggedised laptop costs.

With ~80% of all data deemed to have a spatial component, modern Mobile GIS are a powerful geo-centric business process integration platform enabling the Spatial Enterprise.

Current high level Mobile GIS requirements can be characterised as:

any data	interoperability between proprietary GIS data sources (Smallworld, ESRI, spatial-NET, Open Spatial, ...)
any server	in-house or cloud hosting
any client	iOS, Android, Windows 8 native apps
anyone	ease of use touch client
any time	on- or off-line availability often called 'sometimes connected' mode

Real Business Benefits in Mobile GIS for Utilities and Telcos can be found in:

- Significant Truck Roll reductions
- End-to-end business process flows
- Minimal operator training time, easy-to-use touch client interface
- 100% on/off-line availability
- Better informed and more productive work force
- Complete office-to-field GIS & ERP & DMS integration
- ERP Work Order, Notification, Equipment, Functional Location and Operations
- Integrated Planned & Unplanned Outage Management
- Map-centric business object selection, no database structure knowledge required
- Dynamic data feeds (outages, customer-off-supply, weather, etc...)
- Redlining instantly shared with the office
- Google StreetView & Navigation
- One-click-access to associated documents (work instructions, manuals, OH&S)

Device Limitations

Although not all applications of mobile GIS are limited by the device, many are. These limitations are more applicable to smaller devices such as cell phones and PDAs. Such devices have:

- small screens with a poor resolution
- limited memory and processing power
- a poor (or no) keyboard
- short battery life

Additional limitations can be found in web client based tablet applications:

- poor web GUI and device integration
- on-line reliance
- very limited off-line web client cache

Enterprise GIS

Enterprise GIS refers to a geographical information system that integrates geographic data across multiple departments and serves the whole organisation (ESRI, 2003). The basic idea of an enterprise GIS is to deal with departmental needs collectively instead of individually. When organisations started using GIS in the 1960s and 1970s, the focus was on individual projects where individual users created and maintained data sets on their own desktop computers. Due to extensive interaction and work-flow between departments, many organisations have in recent years switched from independent, stand-alone GIS systems to more integrated approaches that share resources and applications (Ionita, 2006).

Some of the potential benefits that an enterprise GIS can provide include significantly reduced redundancy of data across the system, improved accuracy and integrity of geographic information, and more efficient use and sharing of data (Sipes, 2005). Since data is one of the most significant investments in any GIS program, any approach that reduces acquisition costs while maintaining data quality is important. The implementation of an enterprise GIS may also reduce the overall GIS maintenance and support costs providing a more effective use of departmental GIS resources. Data can be integrated and used in decision making processes across the whole organisation (Sipes, 2005).

Strategy

The development of the European Union (EU) '"INSPIRE"' initiative indicates this is a matter that is gaining more awareness at the national and EU scale. This states that there is a need to create 'quality geo-referenced information' that would be useful for a better understanding of human activities on environmental processes. Therefore, it is an ambitious project that aims to develop a European spatial information database.

The GI strategy for Scotland was introduced in 2005 to provide a sustainable SDI, through the "One Scotland – One Geography" implementation plan. This documentation notes that it should be able to provide linkages to the "Spaces, Faces and Places of Scotland".

Although plans for a GI strategy have been in existence for some time, it was revealed at the AGI Scotland 2007 conference that a recent budget review by the Scottish Government indicated there will not be an allocation of resources to fund this initiative within the next term. Therefore, a business plan will need to be presented in order to outline the cost-benefits involved with taking up the strategy.

Standards

The main standards for Distributed GIS are provided by the Open Geospatial Consortium (OGC). OGC is a non-profit international group which seeks to Web-Enable GIS and in turn Geo-Enable the web. One of the major issues concerning distributed GIS is the interoperability of the data since it can come in different formats using different projection systems. OGC standards seek to provide interoperability between data and to integrate existing data.

OGC

In terms of interoperability, the use of communication standards in Distributed GIS is particularly important. General standards for Geospatial Data have been developed by the Open Geospatial Consortium (OGC). For the exchange of Geospatial Data over the web, the most important OGC standards are Web Map Service (WMS) and Web Feature Service (WFS).

Using OGC compliant gateways allows for building very flexible Distributed GI Systems. Unlike monolithic GI Systems, OGC compliant systems are naturally web-based and do not have strict definitions of servers and clients. For instance, if a user (client) accesses a server, that server itself can act as a client of a number of further servers in order to retrieve data requested by the user. This concept allows for data retrieval from any number of different sources, providing consistent data standards are used.

Furthermore, this concept allows data transfer with systems not capable of GIS functionality. A key function of OGC standards is the integration of different systems already existing and thus geo-enabling the web. Web services providing different functionality can be used simultaneously to combine data from different sources (mash-ups). Thus, different services on distributed servers can be combined for 'service-chaining' in order to add additional value to existing services. Providing a wide use of OGC standards by different web services, sharing distributed data of multiple organisations becomes possible.

Other Standards

Some important languages used in OGC compliant systems are described in the following. XML stands for eXtensible Markup language and is widely used for displaying and interpreting data from computers. Thus the development of a web-based GI system requires several useful XML encodings that can effectively describe two-dimensional graphics such as maps SVG and at the same time store and transfer simple features GML. Because GML and SVG are both XML encodings, it is very straightforward to convert between the two using an XML Style Language Transformation XSLT. This gives an application a means of rendering GML, and in fact is the primary way that it has been accomplished among existing applications today. XML can introduce innovative web services, in terms of GIS. It allows geographic information to be easily translated in graphic and in these terms scalar vector graphics (SVG) can produce high quality dynamic outputs by using data retrieved from spatial databases. In the same aspect Google, one of the pioneers in web-based GIS, has developed its own language which also uses a XML structure. Keyhole Markup Language or KML is a file format used to display geographic data in an earth browser, such as Google Earth, Google Maps, and Google Maps for mobile browsers *"Google KML definition". Retrieved 2007-11-21.*

Global System for Mobile Communications

It is a global standard for mobile phones around the world. Networks using the GSM system offer transmission of voice, data and messages in text and multimedia form and provide web, telenet, ftp, email services etc. over the mobile network. Almost two million people are now using GSM. Five main standards of GSM exist: GSM 400, GSM 850, GSM 900, GSM-1800 (DCS) and GSM1900 (PCS). GSM 850 and GSM 1900 is used in North America, parts of Latin America and parts of Africa. In Europe, Asia and Australia GSM 900/1800 standard is used.

GSM consists of two components: the mobile radio telephone and Subscriber Identity Module. GSM is a cellular network, which is a radio network made up of a number of cells. For each cell, the transmitter (known as a base station) is transmitting and receiving signals. The base station is controlled through the Base Station Controller via the Mobile Switching Centre.

For GSM enhancement GPRS and UMTS technology was introduced. General Packet Radio Service is a packet-oriented data service for data transmission. Universal Mobile Telecommunications System is the Third Generation (3G) mobile communication system. Both provide similar services to 2G, but with greater bandwidth and speed.

Wireless Application Protocol

This is a standard for the data transmission of internet content and services. It is a secure specification that allows users to access the information instantly via mobile phones, pagers, two-way radios, smartphones and communicators. WAP supports HTML and XML, and WML language, and is specifically designed for small screens and one-hand navigation without a keyboard. WML is scalable from two-line text displays up to the graphical screens found on smart phones. It is much stricter than HTML and is similar to JavaScript.

Location-Based Services

Location-based services (LBS) are services that are distributed wirelessly and provide information relevant to the user's current location. These services include such things as 'find my nearest ...', directions, and various vehicle monitoring systems, such as the GM OnStar system amongst others. Location-based services are generally run on mobile phones and PDAs, and are intended for use by the general public more than Mobile GIS systems which are geared towards commercial enterprise. Devices can be located by triangulation using the mobile phone network and/or GPS.

Device limitations are similar to those for mobile GIS. In addition, any devices that are reliant on WAP will be limited by WAP functionality.

Geotagging

Geotagging is the process of adding geographical identification metadata to resources such as websites, RSS feed, images or videos. The metadata usually consist of latitude and longitude coordinates but may also include altitude, camera holding direction, place information and so on. Flickr website is one of the famous web services which host photos and provides functionality to add latitude and longitude information to the picture.

The main idea is to use metadata related to pictures and photo collection. A geotag is simply a properly-formed XML tag giving the geographic coordinates of a place. The coordinates can be specified in latitude and longitude or in UTM (Universal Transverse Mercator) coordinates. The RDFIG Geo vocabulary from the W3C is the common basis for the recommendations. It supplies official global names for the latitude, longitude, and altitude properties. These are given in a system of coordinates known as "the WGS84 datum". (A geographic datum specifies an ellispoidal approximation to the Earth's surface; WGS84 is the most commonly used such datum; it is utilized, e.g. for GPS).

To specify that the longitude of something is X, that its latitude is Y, and, optionally, that its altitude is Z, tags form of the tags used is <geo:long>X</geo:long> <geo:lat>Y</geo:lat> <geo:alt>Z</geo:alt>

Altitude is specified in meters. The prefix "geo:" represents the RDFIG Geo namespace, whose URL is: http://www.w3.org/2003/01/geo/wgs84_pos#.

Geotagging an HTML element

The following tag will pass muster as correct XML in the context of XHTML (the newer dialect of HTML that adheres to the XML standard), but will also work in earlier HTML dialects, in the sense of being tolerated by all modern browsers. To geotag an HTML element, include a span of the following form:

```
<span style="display:none" xmlns:geo="http://www.w3.org/2003/01/
geo/wgs84_pos#">
  <geo:lat>46.1</geo:lat>
  <geo:long>124</geo:long>
 </span>
```

If the geo namespace is defined at an outer level of the document, the namespace definition in the span tag can be omitted, leaving In earlier HTML dialects, omitting the namespace definition is also appropriate, since the objective of adhering to the XML standard is irrelevant. This technique can be used to geotag a post in a weblog, or elements within any HTML document. Geotagging XML (including RSS and RDF).

In XML simply elements of the form

```
<geo:lat>46.1</geo:lat>
<geo:long>124</geo:long>
```

are included as children of the element one wish to tag, and place the definition of the geo namespace at the outermost level of the document. Geotagging a web page Following method is used to assign a location to a web page as a whole, rather than to its parts. In the <head> element, following metatags are included:

```
<meta property="geo:lat">46.1</meta>
<meta property="geo:long">124</meta>
```

In XHTML, the document namespace definition should include the geo tag. This form of meta tag follows the recommendations contained in http://www.w3.org/MarkUp/2004/02/xhtml-rdf.html

Mashups

In distributed GIS, the term mashup refers to a generic web service which combines content and functionality from disparate sources; mashups reflect a separation of information and presentation. Mashups are increasingly being used in commercial and government applications as well as in the public domain.

When used in GIS, it reflects the concept of connecting your application with a mapping service (e.g., combining Google maps with Chicago crime statistics to create the [www.chicagocrime.org/map/ Chicago crime statistics map]).

Mashups are fast, provide value for money and remove responsibility for the data from the creator.

Second generation systems provide mashups mainly based on URL parameters, while Third generation systems (e.g. Google Maps) allow customisation via script (e.g. JavaScript).

Web Mapping Services

A web mapping service is a means of displaying and interacting with maps on the Web. The first web mapping service was the Xerox PARC Map Viewer built in 1993 and decommissioned in 2000.

There have been 3 generations of web map service. The first generation was from 1993 onwards and consisted of simple image maps which had a single click function. The second generation was from 1996 onwards and still used image maps the one click function. However, they also had zoom and pan capabilities (although slow) and could be customised through the use of the URL API. The third generation was from 1998 onwards and were the first to include slippy maps. They utilise AJAX technology which enables seamless panning and zooming. They are customisable using the URL API and can have extended functionality programmed in using the DOM.

Web map services are based on the concept of the image map whereby this defines the area overlaying an image (e.g. GIF). An image map can be processed client or server side. As functionality is built into the web server, performance is good. Image maps can be dynamic. When image maps are used for geographic purposes, the co-ordinate system must be transformed to the geographical origin to conform to the geographical standard of having the origin at the bottom left corner.

Web maps are used for location-based services.

Examples of web mapping services are:

- Streetmaps
- Google Maps
- Multimap

Web 2.0

The Internet has gone far beyond the average persons thinking. It is not only chatting or just surfing the web but it is a new way of connecting people and how to bring the experience from a desktop to the browser which will be more user friendly. For this reason some Rich Internet Applications (RIA) are required. Ajax is one of the widely used terms in this context in conjunction with flash, flex and Nexaweb. Web 2.0 application tends to interact with the end user and the end user had a greater role in the web 2.0 applications as he/she is not only the user of the application but also a participant. This may be through tagging the content, contributing to wiki, by podcasting or blogging.

The user also provides feedback on the applications in addition to their social contribution to the applications.

One of the key components of web 2.0 are web services, and how these applications expose their functionality and can be combined with other applications to provide a rich set of new applications using mashups. Computer languages are required to perform these tasks, and it is very important to update the applications frequently so the many users get up to date information. Some of the applications of web 2.0 are flickr, del.icio.us, YouTube, Facebook, skyligo and Myspace.

Performance

The speedup of a program as a result of parallelization is given by Amdahl's law. Amdahl's Law states that potential program speedup is defined by the fraction of code (P) that can be parallelized: 1/(1-P)

If the code cannot be broken up to run over multiple processors, P = 0 and the speedup = 1 (no speedup). If it is possible to break up the code to be perfectly parallel then P = 1 and the speedup is infinite (in theory, although other factors such as scalability and complexity limit this possibility). Thus, there is an upper bound on the usefulness of adding more parallel execution units.

Gustafson's law is a law closely related to Amdahl's law but doesn't make as many assumptions and tries to model these factors in the representation of performance. The equation can be modelled by $S(P) = P - \alpha * (P - 1)$ where P is the number of processors, S is the speedup, and α the non-parallelizable part of the process.

Parallel Processing

Parallel processing is the use of multiple CPU's to execute different sections of a program together. Remote sensing and surveying equipment have been providing vast amounts of spatial information, and how to manage, process or dispose of this data have become major issues in the field of Geographic Information Science (GIS).

To solve these problems there has been much research into the area of parallel processing of GIS information. This involves the utilization of a single computer with multiple processors or multiple computers that are connected over a network working on the same task. There are many different types of distributed computing, two of the most common are clustering and grid processing.

Why Us2e Parallel Processing

- The primary reasons for using parallel computing:
    - Save time - wall clock time

- Solve larger problems
- Provide concurrency (do multiple things at the same time)

- Other reasons might include:
  - Taking advantage of non-local resources - using available computing resources on a wide area network, or even the Internet when local computing resources are scarce.
  - Cost savings - using multiple "cheap" computing resources instead of paying for time on a supercomputer.
  - Overcoming memory constraints - single computers have very finite memory resources. For large problems, using the memories of multiple computers may overcome this obstacle.
- Limits to serial computing - both physical and practical reasons pose significant constraints to simply building ever faster serial computers:
  - Transmission speeds - the speed of a serial computer is directly dependent upon how fast data can move through hardware. Absolute limits are the speed of light (30 cm/nanosecond) and the transmission limit of copper wire (9 cm/nanosecond). Increasing speeds necessitate increasing proximity of processing elements.
  - Limits to miniaturization - processor technology is allowing an increasing number of transistors to be placed on a chip. However, even with molecular or atomic-level components, a limit will be reached on how small components can be.
  - Economic limitations - it is increasingly expensive to make a single processor faster. Using a larger number of moderately fast commodity processors to achieve the same (or better) performance is less expensive.
- The future: during the past 10 years, the trends indicated by ever faster networks, distributed systems, and multi-processor computer architectures (even at the desktop level) clearly show that parallelism is the future of computing.

Grid Computing

Some consider this to be "the third information technology wave" after the Internet and Web, and will be the backbone of the next generation of services and applications that are going to further the research and development of GIS and related areas.

Grid computing allows for the sharing of processing power, enabling the attainment of high performances in computing, management and services. Grid computing, (unlike the conventional supercomputer that does parallel computing by linking multiple processors over a system bus) uses a network of computers to execute a program. The problem of using multiple computers lies in the difficulty of dividing up the tasks among the computers, without having to reference portions of the code being executed on other CPUs.

Local Search

Local Search is a recent approach to internet searching that incorporates geographical information into search queries so that the links that you return are more relevant to where you are. It developed out of an increasing awareness that many search engine users are using it to look for a business or service in the local area. Local search has stimulated the development of web mapping, which is used either as a tool to use in geographically restricting your search or as an additional resource to be returned along with search result listings. It has also led to an increase in the number of small businesses advertising on the web.

Distributed GIS – Acronym Index

AIS Automatic Identification System
CAT Catalogue Service
CID Complete Intervisibility Database
DEG Display Element Generator
GSDI (Geo)Spatial Data Infrastructure
NGDF National Geospatial Data Framework
NSDI National and International Spatial Data Infrastructures
OGDI Open Geographic/Geospatial Datastore Interface
OGSA Open Grid Services Architecture
SDI Spatial Data Infrastructure
SRS Spatial Reference Systems
WMC Web Map Context

Location Intelligence

Location intelligence (LI), or spatial intelligence, is the process of deriving meaningful insight from geospatial data relationships to solve a particular problem. It involves layering multiple data sets spatially and/or chronologically, for easy reference on a map, and its applications span industries, categories and organizations It is generally agreed that more than 80% of all data has a location element to it and that location directly affects the kinds of insights that you might draw from many sets of information. Maps have been used to represent information throughout the ages, but what might be referenced as the first example of true location 'intelligence' was in London in 1854 when

John Snow was able to debunk theories about the spread of cholera by overlaying a map of the area with the location of water pumps and was able to narrow the source to a single water pump. This layering of information over a map was able to identify relationships, and in turn insights that might otherwise never have been understood. This is the core of location intelligence today.

Deploying location intelligence by analyzing data using a geographical information system (GIS) within business is becoming a critical core strategy for success in an increasingly competitive global economy. Location or GIS tools enable spatial experts to collect, store, analyze and visualize data. Location intelligence experts are defined by their advanced education in spatial technology and applied use of spatial methodologies.

Location intelligence experts can use a variety of spatial and business analytical tools to measure optimal locations for operating a business or providing a service. Location intelligence experts begin with defining the business ecosystem which has many interconnected economic influences. Such economic influences include but are not limited to culture, lifestyle, labor, healthcare, cost of living, crime, economic climate and education.

The term "location intelligence" is often used to describe the people, data and technology employed to geographically "map" information. These mapping applications can transform large amounts of data into color-coded visual representations that make it easy to see trends and generate meaningful intelligence. The creation of location intelligence is directed by domain knowledge, formal frameworks, and a focus on decision support. Location cuts across through everything i.e. devices, platforms, software and apps, and is one of the most important ingredient of understanding context in sync with social data, mobile data, user data, sensor data, using platforms as CartoDB where data as a service and the analytical and visualisation tools blend together to create a business friendly environment.

Location intelligence is also used to describe the integration of a geographical component into business intelligence processes and tools, often incorporating spatial database and spatial OLAP tools.

In 2012, Wayne Gearey from the commercial real estate industry was selected to offer the first applied course on location intelligence at the University of Texas at Dallas. In this course, Gearey defines location intelligence as the process for selecting the optimal location that will support workplace success and address a variety of business and financial objectives.

Geoblink defines location intelligence as the capability to understand and optimize a physical network of points of sale in the process of making business decisions.

Pitney Bowes MapInfo Corporation describes location intelligence as follows: "Spatial information, commonly known as "Location", relates to involving, or having the nature of where. Spatial is not constrained to a geographic location however most common business uses o spatial information deal with how spatial information is tied to a location on the earth. Miriam-Webster® defines Intelligence as "The ability to learn or understand, or the ability to apply knowledge to manipulate one`s environment." Combining these terms alludes to how you achieve an understanding of the spatial aspect of information and apply it to achieve a significant competitive advantage."

Definition by ESRI is as follows: "Location Intelligence is defined as the capacity to organize and understand complex data through the use of geographic relationships. LI organizes business and geographically referenced data to reveal the relationship of location to people, events, transactions, facilities, and assets."

Definition by Yankee Group within their White Paper "Location Intelligence in Retail Banking: "...a business management term that refers to spatial data visualization, contextualization and analytical capabilities applied to solve a business problem."

Commercial Applications

Today, location intelligence is used by a broad range of industries to improve overall business results. Applications include:

- Communications & telecommunications: Network planning and design, boundary identification, identifying new customer markets.
- Financial services: Optimize branch locations, market analysis, share of wallet and cross-sell activities, mergers & acquisitions, industry sector analysis, risk management.
- Government: Census updates, law enforcement crime analysis, emergency response, environmental and land management, electoral redistricting, tax jurisdiction assignment, urban planning.
- Healthcare: Site selection, market segmentation, network analysis, growth assessments.
- Higher education: Student Recruitment, Alumni & Donor Tracking, Campus Mapping.
- Hotels & restaurants: Customer profile analysis, site selection, target marketing, expansion planning.
- Insurance: Address validation, underwriting and risk management, claims management, marketing and sales analysis, market penetration studies.
- Media: Target market identification, subscriber demographics, media planning.

- Realestate: Site reports, comprehensive site analysis, retail modeling, presentation quality maps.
- Retail: Site selection, maximize per-store sales, identify under-performing stores, market analysis.
- Transportation: Transport planning, route monitoring.
- K-12 : School Site Selection, enrollment planning, school attendance area modification (boundary change), school consolidation, district consolidation, student achievement plotting.

Geographic Information Systems in Geospatial Intelligence

Geographic information systems (GIS) play a constantly evolving role in geospatial intelligence (GEOINT) and United States national security. These technologies allow a user to efficiently manage, analyze, and produce geospatial data, to combine GEOINT with other forms of intelligence collection, and to perform highly developed analysis and visual production of geospatial data. Therefore, GIS produces up-to-date and more reliable GEOINT to reduce uncertainty for a decisionmaker. Since GIS programs are Web-enabled, a user can constantly work with a decision maker to solve their GEOINT and national security related problems from anywhere in the world. There are many types of GIS software used in GEOINT and national security, such as Google Earth, ERDAS IMAGINE, GeoNetwork opensource, and Esri ArcGIS.

Background

Geographic Information Systems (GIS)

A user can enter different kinds of data in map form into a GIS to begin their analysis, such as United States Geological Survey (USGS) digital line graph data, contour lines, elevation maps, topographic maps, geologic maps, and satellite imagery. A user can also convert digital information into forms that a GIS can identify and utilize, such as census tabular data or Microsoft Excel files. Users can easily capture digital data in a GIS. If the data is not digital, then users will need to employ various techniques to capture the data, such as digitizing maps by hand-tracing with a computer mouse, utilizing a digitizing tablet to collect feature coordinates, using electronic scanners, or uploading Global Positioning System (GPS) coordinates.

GIS applies to the geographical facets of various aspects of everyday life, such as transportation, logistics, medicine, marketing, sociology, ecology, pure and applied sciences, emergency management, and criminology. GIS is also utilized in all three areas of intelligence: national security intelligence, law enforcement intelligence, and competitive intelligence

Geospatial Intelligence (GEOINT)

GEOINT, known previously as imagery intelligence (IMINT), is an intelligence collection discipline that applies to national security intelligence, law enforcement intelligence, and competitive intelligence. For example, an analyst can use GEOINT to identify the route of least resistance for a military force in a hostile country, to discover a pattern in the locations of reported burglaries in a neighborhood, or to generate a map and comparison of failing businesses that a company is likely to purchase. GEOINT is also the geospatial product of a process that is focused externally, designed to reduce the level of uncertainty for a decisionmaker, and that uses information derived from all sources. The National Geospatial-Intelligence Agency (NGA), who has overall responsibility for GEOINT in the U.S. Intelligence Community (IC), defines GEOINT as "information about any object—natural or man-made—that can be observed or referenced to the Earth, and has national security implications."

Some of the sources of collected imagery information for GEOINT are imagery satellites, cameras on airplanes, Unmanned Aerial Vehicles (UAV) and drones, handheld cameras, maps, or GPS coordinates. Recently the NGA and IC have increased the use of commercial satellite imagery for intelligence support, such as the use of the IKONOS, Landsat, or SPOT satellites. These sources produce digital imagery via electro-optical systems, radar, infrared, visible light, multispectral, or hyperspectral imageries.

The advantages of GEOINT are that imagery is easily consumable and understood by a decisionmaker, has low human life risk, displays the capabilities of a target and its geographical relationship to other objects, and that analysts can use imagery world-wide in a short time. On the other hand, the disadvantages of GEOINT are that imagery is only a snapshot of a moment in time, can be too compelling and lead to ill-informed decisions that ignore other intelligence, is static and vulnerable to deception and decoys, does not depict the intentions of a target, and is expensive and subject to environmental problems.

GIS use in GEOINT and National Security Intelligence

Overview

A majority of national security intelligence decisions involve geography and GEOINT. GIS allows the user to capture, manage, exploit, analyze, and visualize geographically referenced information, physical features, and other geospatial data. GIS is thus a critical infrastructure for the GEOINT and national security community in manipulating and interpreting spatial knowledge in an information system. GIS extracts real world geographic or other information into datasets, maps, metadata, data models, and workflow models within a geodatabase that is used to solve GEOINT-related problems. GIS provides a structure for map and data production that allows a user to add other data sources, such as satellite or UAV imagery, as new layers to a geodatabase. The geodatabase can be disseminated and operated across any network of associated users (i.e. from the GEOINT analyst to the warfighter) and engenders a common spatial capability for all defense and intelligence domains.

The map and chart production agency and imagery intelligence agency, the principal two agencies of GEOINT, use GIS to efficiently work together to solve decisionmaker's geospatial questions, to communicate effectively between their unique departments, and to provide constantly updated, accurate GEOINT to their national security and warfighter domains.

Another important aspect of GIS is its ability to fuse geospatial data with other forms of intelligence collection, such as signals intelligence (SIGINT), measurement and signature intelligence (MASINT), human intelligence (HUMINT), or open source intelligence (OSINT). A GIS user can incorporate and fuse all of these types of intelligence into applications that provide corroborated GEOINT throughout an organization's information system.

GIS enables efficient management of geospatial data, the fusion of geospatial data with other forms of intelligence collection, and advanced analysis and visual production of geospatial data. This produces faster, corroborated, and more reliable GEOINT that aims to reduce uncertainty for a decisionmaker.

Roles

- Data and map production
- Data fusion, data discovery through metadata catalogs, and data dissemination through Web portals and browsers
- Analysis and exploitation of collected imagery or intelligence
- SIGINT, GEOINT, MASINT, and other sensor analysis
- Fusion of multiple forms of intelligence collection
- Collaborative planning and efficient workflow management between decisionmakers, analysts, consumers, and warfighters
- Suitability and temporal analysis
- Stewardship: Geospatial intelligence

Related Esri Products

Distributed Geospatial Intelligence Network (DGInet)

The DGInet technology allows military and national security intelligence customers to access large multi-terabyte databases through a common Web-based interface. This gives the users the capability to quickly and easily identify, overlay, and fuse georeferenced data from various sources to create maps or support geospatial analysis. Esri designed the technology for inexperienced GIS users of national security intelligence and defense organizations in order to provide a Web-based enterprise solution for publishing, distributing, and exploiting GEOINT data among designated organizations. According to Esri, the DGInet technology "uses thin clients to search massive amounts of geospatial and intelligence data using low-bandwidth Web services for data discovery, dissemination, and horizontal fusion of data and products."

PLTS for ArcGIS Specialized Solutions

PLTS for ArcGIS Specialized Solutions is a group of software applications that extends ArcGIS to facilitate database driven cartographic production for geospatial and mapping agencies, nautical and aeronautical chart production, foundation mapping, and defense mapping requirements. The collection of software applications includes Esri Production Mapping, Esri Nautical Solution, Esri Aeronautical Solution, and Esri Defense Mapping programs that provide quality control, easier and consistent map production, database sharing, and efficient workflow management for each program's specific type of mapping or charting.

Geoprocessing

Geoprocessing is based on a framework of data transformation in GIS and is a collection of hundreds of GIS tools that manipulate geospatial or other data in GIS. A geoprocessing tool performs an operation (often the name of the tool, such as "Clip") on an existing GIS dataset and produces a new dataset as a result of the utilized tool. GIS users utilize these tools to create a workflow model that quickly and easily transforms raw data into the desired product.

In GEOINT, users employ geoprocessing in similar ways. They can make geoprocessing tools resemble analytic techniques to transform large amounts of data into actionable information. In national security intelligence and defense organizations, geoprocessing notifies users to events occurring in specific areas of interest and enables domain-specific analysis applications, such as radio frequency analysis, terrain analysis, and network analysis.

Tracking Analyst and Tracking Server

The ArcGIS Tracking Analyst extension enables the user to create time series visualizations to analyze time and location sensitive information. It creates a visible path from incorporated data that shows movement through space and time. The program allows the national security intelligence or defense user to track assets (such as vehicles or personnel), monitor sensors, visualize change over time, play back events, and analyze historical or real-time temporal data.

The Tracking Server program is an Esri enterprise technology that integrates real-time data with GIS to disseminate information quickly and easily to decisionmakers. This program enables the user to obtain data in any format and transmit it to the necessary consumer or ArcGIS Tracking Analyst user, to conduct filters or alerts on specific attributes of incoming data or global positions, and to log data into ArcGIS Server for efficient project management and information sharing.

When Tracking Server and ArcGIS Tracking Analyst are used together, a user can monitor changes in data as they occur in real-time. A national security intelligence or defense user can subscribe to real-time data over the Internet from GPS and custom data feeds to support GEOINT requirements, such as fleet management or target tracking.

ArcGIS Military Analyst

The ArcGIS Military Analyst extension incorporates display and analysis tools that allow the use and production of vector and raster products, line-of-sight analysis, hillshade analysis, terrain analysis, and Military Grid Reference System (MGRS) conversion. This program also provides a basis for command, control, and intelligence (C2I) systems. National security intelligence and defense organizations use ArcGIS Military Analyst extension to integrate geospatial data with other defense data, analyze digital terrains, and prepare for battle. This program also enables such users to manage and analyze geospatial data and relationships between mission planning, logistics, and C2I.

Military Overlay Editor (MOLE)

MOLE is a set of command components that enables national security intelligence and defense users to easily create, display, and edit U.S. Department of Defense MIL-STD-2525B and the North Atlantic Treaty Organization APP-6A military symbology in a map. This allows for easier and faster identification, understanding, and movement of ally and hostile forces on a map by combining GIS spatial analysis techniques with common military symbols. MOLE provides a clearer visualization of mission planning and goals for the decisionmaker, and allows a user to import, locate, and display order of battle databases.

Grid Manager

Grid Manager enables the national security intelligence or defense user to create accurate, realistic grids that contain geographic location indicators based on specified shapes, scales, coordinate systems, and units. This program allows the user to create multiple grids, graticules, and borders for such map products as MGRS coordinates and tourist, topographic, parcel, street, nautical, and aeronautical maps.

GIS use in the National Geospatial-Intelligence Agency (NGA)

The NGA uses GIS products to create digital nautical, aeronautical, and topographic charts and maps, to perform geotechnical and coordinate system analysis, and to help solve a large variety of national security and military problems. Since the NGA is a U.S. Department of Defense combat support agency and a member of the IC, it uses GIS to produce precise, up-to-date GEOINT for members of the U.S. Armed Forces, the IC, and other government agencies. Web-enabled GIS applications allow for fast, efficient sharing and disseminating of geospatial data, products, and intelligence from the NGA to its allies, warfighters, partners, and other agencies across the World Wide Web. The NGA and Esri have successfully collaborated on providing timely, accurate, and relevant GEOINT in support of U.S. national security for the past 20 years.

The NGA has created a grouping of web-based capabilities called GEOINT Online. This program allows a user to search and access all NGA GEOINT documents from wherever they are stored and from wherever the user is. GEOINT Online provides quick, easy, and reliable access to current NGA intelligence products, changes in activities or regions, information from analyst's blogs and Intellipedia, geospatial imagery, maps and charts, major GIS commercial software packages, and GIS combinations of these products. A user can also edit and format existing NGA/GIS products and maps to create, print, and download new products that fulfill current decisionmaker requirements. Ultimately, this results in the faster production of timely and relevant GEOINT data. This program allowed the NGA to change its focus from simply generating cartographic products to providing updated, accurate GEOINT to support the national security and military requirements of its customers.

References

- George Moon (2008-00-00). "Location Intelligence – Meeting IT Expectation" (PDF). Pitney Bowes. Retrieved 2015-10-05.
- Marcus Torchia (2009-00-00). "Location Intelligence in Retail Banking" (PDF). Pitney Bowes. Retrieved 2015-10-05.
- "Arc Digitized Raster Graphic (ADRG)". Digital Preservation. Library of Congress. 2011-09-25. Retrieved 2014-03-13.
- Geographic Information Systems (GIS) Poster, Last modified on 2007-02-22, USGS, Retrieved on 2011-01-16

3

Tools and Techniques of Geographic Information Systems

Tools and techniques are an important component of any field of study. The tools and techniques discussed within this content are ichthyology and GIS, spatial analysis, geo-tagged photography, remote sensing and rubbersheeting. A GIS can provide accurate data of underwater geography whereas spatial analysis is the technique used to study geographical properties. The following chapter helps the reader in developing an in-depth understanding of the techniques used in geographic information systems.

Ichthyology and GIS

A Geographic Information System is a tool for mapping and analyzing data. The ability to layer many features onto the same map and select or unselect as needed allows for a multitude of views and ease of interpreting data. More important, this allows for in depth scientific analysis and problem solving.

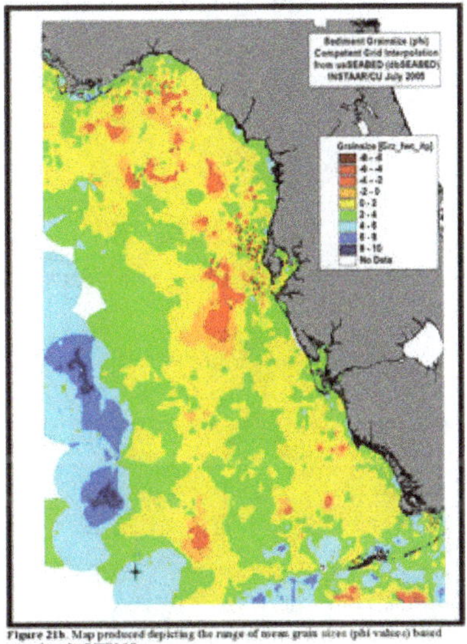

Figure 21b. Map produced depicting the range of mean grain sizes (phi values) based on analyses by INSTAAR.

Mean grain sizes found as sediment.

Ichthyology involves many areas of study related to fishes and their habitat. The natural habitat is water, but fish are dependent upon many other factors. Water quality, type, food, cover, sediment are essential for the life cycle of any given fish. Being able to map the presence of certain species with layers of these features provides invaluable insight into species requirements. GIS is an essential tool that allows immediate visualization of all data present and to accurately interpret impacts of habitat degradation or species success.

GIS

GIS is useful when data is specific to a location. It is used to classify, analyze and understand data relationships based on the location and then drawing conclusions from the data. Data capture can occur in the field on small, handheld GPS devices, and then imported and compared to an existing map. This freedom of movement between field and computer is critical to streamlining data collection in field endeavors and generating more accurate data sets.

Ichthyology

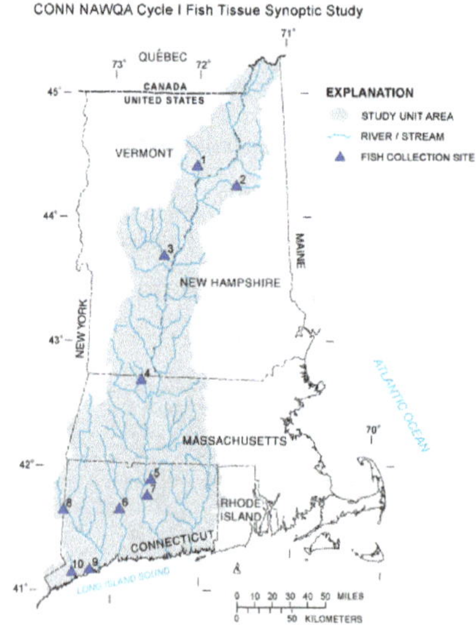

Connecticut fish sampling sites on water.

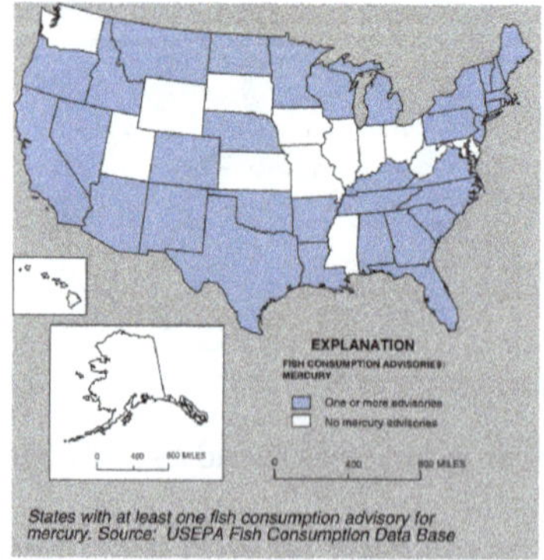

Mercury in fish consumption adivsory.

Ichthyology requires an understanding species geographic requirements. Fish require different abiotic environments or sediments for successful completion of biological life cycle based on species. Serious examinations of species should always include habitat because habitat differences create changes in population. Sediment could thereby be mapped and changes in sediment could easily be verified using previous records while simultaneously showing changes in resident fish populations. Various factors relating to the fish life cycle, such as food sources, migration patterns, changes in spawning grounds, could all be more accurately explained and documented using GIS versus a more traditional paper. More important, data could be gathered in the field on hand-held GPS units and downloaded directly to an existing map. Streamlining data entry removes error by having observations made and recorded and entered in the field while observations are actively being recorded and uploading the data to a computer upon return to the lab. The other alternative is making observations in the field and then recording the data upon return to the lab; this second technique can allow for opinions to affect how data in interpreted during collection.

Advantages

One of the biggest advantages in using GIS is assimilating information and using it to highlight significance or irrelevant data. Use of GIS increases the possible integration of many different types of data into a single usable resource making analysis and interpretation easier as well as increasing management of the data involved. Ichthyology is a field of study that requires active examination of a multitude of areas at the same time for accurate study. GIS programs improve spatial data aspects frequently to accurately represent substrate, habitat, quality or other various factors. This tool enables scientists to access information and share results quickly and concisely. Scientists realize the need for a GIS component in their research as evidenced by the founding of such groups as Fishery-Aquatic Research Group. This group organizes an annual symposium with the expressed goal of furthering ichthyology research utilizing GIS.

Spatial Analysis

Spatial analysis or spatial statistics includes any of the formal techniques which study entities using their topological, geometric, or geographic properties. Spatial analysis includes a variety of techniques, many still in their early development, using different analytic approaches and applied in fields as diverse as astronomy, with its studies of the placement of galaxies in the cosmos, to chip fabrication engineering, with its use of "place and route" algorithms to build complex wiring structures. In a more restricted sense, spatial analysis is the technique applied to structures at the human scale, most notably in the analysis of geographic data.

Complex issues arise in spatial analysis, many of which are neither clearly defined nor completely resolved, but form the basis for current research. The most fundamental of these is the problem of defining the spatial location of the entities being studied.

Classification of the techniques of spatial analysis is difficult because of the large number of different fields of research involved, the different fundamental approaches which can be chosen, and the many forms the data can take.

History of Spatial Analysis

Spatial analysis can perhaps be considered to have arisen with early attempts at cartography and surveying but many fields have contributed to its rise in modern form. Biology contributed through botanical studies of global plant distributions and local plant locations, ethological studies of animal movement, landscape ecological studies of vegetation blocks, ecological studies of spatial population dynamics, and the study of biogeography. Epidemiology contributed with early work on disease mapping, notably John Snow's work of mapping an outbreak of cholera, with research on mapping the spread of disease and with location studies for health care delivery. Statistics has contributed greatly through work in spatial statistics. Economics has contributed notably through spatial econometrics. Geographic information system is currently a major contributor due to the importance of geographic software in the modern analytic toolbox. Remote sensing has contributed extensively in morphometric and clustering analysis. Computer science has contributed extensively through the study of algorithms, notably in computational geometry. Mathematics continues to provide the fundamental tools for analysis and to reveal the complexity of the spatial realm, for example, with recent work on fractals and scale invariance. Scientific modelling provides a useful framework for new approaches.

Fundamental Issues in Spatial Analysis

Spatial analysis confronts many fundamental issues in the definition of its objects of study, in the construction of the analytic operations to be used, in the use of computers for analysis, in the limitations and particularities of the analyses which are known, and in the presentation of analytic results. Many of these issues are active subjects of modern research.

Common errors often arise in spatial analysis, some due to the mathematics of space, some due to the particular ways data are presented spatially, some due to the tools which are available. Census data, because it protects individual privacy by aggregating data into local units, raises a number of statistical issues. The fractal nature of coastline makes precise measurements of its length difficult if not impossible. A computer software fitting straight lines to the curve of a coastline, can easily calculate the lengths of the lines which it defines. However these straight lines may have no inherent meaning in the real world, as was shown for the coastline of Britain.

These problems represent a challenge in spatial analysis because of the power of maps as media of presentation. When results are presented as maps, the presentation combines spatial data which are generally accurate with analytic results which may be inaccurate, leading to an impression that analytic results are more accurate than the data would indicate.

Spatial Characterization

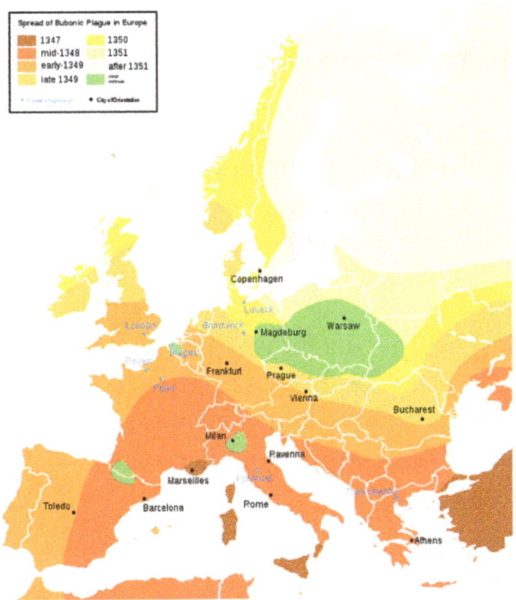

Spread of bubonic plague in medieval Europe. The colors indicate the spatial distribution of plague outbreaks over time. Possibly due to the limitations of printing or for a host of other reasons, the cartographer selected a discrete number of colors to characterize (and simplify) reality.

The definition of the spatial presence of an entity constrains the possible analysis which can be applied to that entity and influences the final conclusions that can be reached. While this property is fundamentally true of all analysis, it is particularly important in spatial analysis because the tools to define and study entities favor specific characterizations of the entities being studied. Statistical techniques favor the spatial definition of objects as points because there are very few statistical techniques which operate directly on line, area, or volume elements. Computer tools favor the spatial definition of objects as homogeneous and separate elements because of the limited number of database elements and computational structures available, and the ease with which these primitive structures can be created.

Spatial Dependency or Auto-correlation

Spatial dependency is the co-variation of properties within geographic space: characteristics at proximal locations appear to be correlated, either positively or negatively. Spatial dependency leads to the spatial autocorrelation problem in statistics since, like temporal autocorrelation, this violates standard statistical techniques that assume independence among observations. For example, regression analyses that do not compensate for spatial dependency can have unstable parameter estimates and yield unreliable significance tests. Spatial regression models capture these relationships and do not suffer from these weaknesses. It is also appropriate to view spatial dependency as a source of information rather than something to be corrected.

Locational effects also manifest as spatial heterogeneity, or the apparent variation in a process with respect to location in geographic space. Unless a space is uniform and boundless, every location will have some degree of uniqueness relative to the other locations. This affects the spatial dependency relations and therefore the spatial process. Spatial heterogeneity means that overall parameters estimated for the entire system may not adequately describe the process at any given location.

Scaling

Spatial measurement scale is a persistent issue in spatial analysis; more detail is available at the modifiable areal unit problem (MAUP) topic entry. Landscape ecologists developed a series of scale invariant metrics for aspects of ecology that are fractal in nature. In more general terms, no scale independent method of analysis is widely agreed upon for spatial statistics.

Sampling

Spatial sampling involves determining a limited number of locations in geographic space for faithfully measuring phenomena that are subject to dependency and heterogeneity. Dependency suggests that since one location can predict the value of another location, we do not need observations in both places. But heterogeneity suggests that this relation can change across space, and therefore we cannot trust an observed degree of dependency beyond a region that may be small. Basic spatial sampling schemes include random, clustered and systematic. These basic schemes can be applied at multiple levels in a designated spatial hierarchy (e.g., urban area, city, neighborhood). It is also possible to exploit ancillary data, for example, using property values as a guide in a spatial sampling scheme to measure educational attainment and income. Spatial models such as autocorrelation statistics, regression and interpolation can also dictate sample design.

Common Errors in Spatial Analysis

The fundamental issues in spatial analysis lead to numerous problems in analysis including bias, distortion and outright errors in the conclusions reached. These issues are often interlinked but various attempts have been made to separate out particular issues from each other.

Length

In a paper by Benoit Mandelbrot on the coastline of Britain it was shown that it is inherently nonsensical to discuss certain spatial concepts despite an inherent presumption of the validity of the concept. Lengths in ecology depend directly on the scale at which they are measured and experienced. So while surveyors commonly measure the length of a river, this length only has meaning in the context of the relevance of the measuring technique to the question under study.

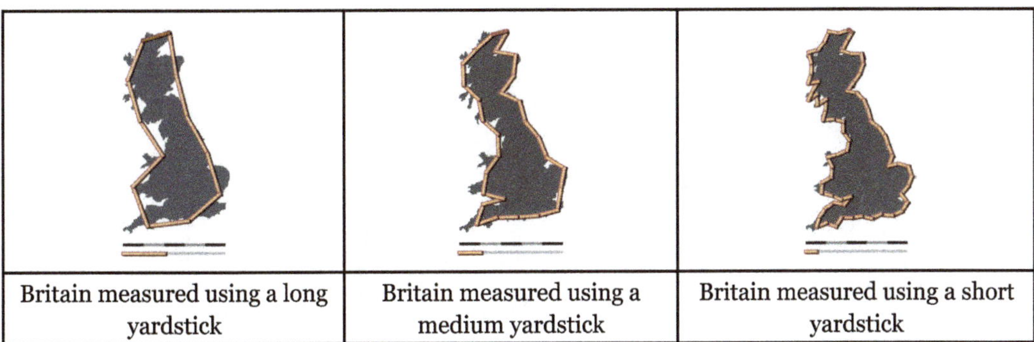

Britain measured using a long yardstick	Britain measured using a medium yardstick	Britain measured using a short yardstick

Locational Fallacy

The locational fallacy refers to error due to the particular spatial characterization chosen for the elements of study, in particular choice of placement for the spatial presence of the element.

Spatial characterizations may be simplistic or even wrong. Studies of humans often reduce the spatial existence of humans to a single point, for instance their home address. This can easily lead to poor analysis, for example, when considering disease transmission which can happen at work or at school and therefore far from the home.

The spatial characterization may implicitly limit the subject of study. For example, the spatial analysis of crime data has recently become popular but these studies can only describe the particular kinds of crime which can be described spatially. This leads to many maps of assault but not to any maps of embezzlement with political consequences in the conceptualization of crime and the design of policies to address the issue.

Atomic Fallacy

This describes errors due to treating elements as separate 'atoms' outside of their spatial context. The fallacy is about transferring individual conclusions to spatial units.

Ecological Fallacy

The ecological fallacy describes errors due to performing analyses on aggregate data when trying to reach conclusions on the individual units. Errors occur in part from spatial aggregation. For example, a pixel represents the average surface temperatures within an area. Ecological fallacy would be to assume that all points within the area have the same temperature. This topic is closely related to the modifiable areal unit problem.

Solutions to the Fundamental Issues

Geographic Space

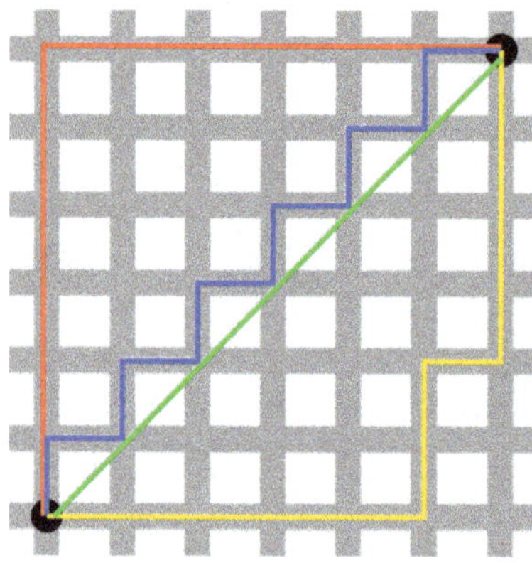

Manhattan distance versus Euclidean distance: The red, blue, and yellow lines have the same length (12) in both Euclidean and taxicab geometry. In Euclidean geometry, the green line has length $6\times\sqrt{2} \approx 8.48$, and is the unique shortest path. In taxicab geometry, the green line's length is still 12, making it no shorter than any other path shown.

A mathematical space exists whenever we have a set of observations and quantitative measures of their attributes. For example, we can represent individuals' incomes or years of education within a coordinate system where the location of each individual can be specified with respect to both dimensions. The distances between individuals within this space is a quantitative measure of their differences with respect to income and education. However, in spatial analysis we are concerned with specific types of mathematical spaces, namely, geographic space. In geographic space, the observations correspond to locations in a spatial measurement framework that captures their proximity in the real world. The locations in a spatial measurement framework often represent locations on the surface of the Earth, but this is not strictly necessary. A spatial measurement framework can also capture proximity with respect to, say, interstellar space or within a biological entity such as a liver. The fundamental tenet is Tobler's First Law of Geography: if the interrelation between entities increases with proximity in the real world, then representation in geographic space and assessment using spatial analysis techniques are appropriate.

The Euclidean distance between locations often represents their proximity, although this is only one possibility. There are an infinite number of distances in addition to Euclidean that can support quantitative analysis. For example, "Manhattan" (or "Taxicab") distances where movement is restricted to paths parallel to the axes can be more meaningful than Euclidean distances in urban settings. In addition to distances, other geographic relationships such as connectivity (e.g., the existence or degree of shared borders) and direction can also influence the relationships among entities. It is also possible to compute minimal cost paths across a cost surface; for example, this can represent proximity among locations when travel must occur across rugged terrain.

Types of Spatial Analysis

Spatial data comes in many varieties and it is not easy to arrive at a system of classification that is simultaneously exclusive, exhaustive, imaginative, and satisfying. -- G. Upton & B. Fingelton

Spatial Data Analysis

Urban and Regional Studies deal with large tables of spatial data obtained from censuses and surveys. It is necessary to simplify the huge amount of detailed information in order to extract the main trends. Multivariable analysis (or Factor analysis, FA) allows a change of variables, transforming the many variables of the census, usually correlated between themselves, into fewer independent "Factors" or "Principal Components" which are, actually, the eigenvectors of the data correlation matrix weighted by the inverse of their eigenvalues. This change of variables has two main advantages:

1. Since information is concentrated on the first new factors, it is possible to keep only a few of them while losing only a small amount of information; mapping them produces fewer and more significant maps

2. The factors, actually the eigenvectors, are orthogonal by construction, i.e. not correlated. In most cases, the dominant factor (with the largest eigenvalue) is the Social Component, separating rich and poor in the city. Since factors are not-correlated, other smaller processes than social status, which would have remained hidden otherwise, appear on the second, third, ... factors.

Factor analysis depends on measuring distances between observations : the choice of a significant metric is crucial. The Euclidean metric (Principal Component Analysis), the Chi-Square distance (Correspondence Analysis) or the Generalized Mahalanobis distance (Discriminant Analysis) are among the more widely used. More complicated models, using communalities or rotations have been proposed.

Using multivariate methods in spatial analysis began really in the 1950s (although some examples go back to the beginning of the century) and culminated in the 1970s, with the increasing power and accessibility of computers. Already in 1948, in a seminal publication, two sociologists, Bell and Shevky, had shown that most city populations in the USA and in the world could be represented with three independent factors : 1- the « socio-economic status » opposing rich and poor districts and distributed in sectors running along highways from the city center, 2- the « life cycle », i.e. the age structure of households, distributed in concentric circles, and 3- « race and ethnicity », identifying patches of migrants located within the city. In 1961, in a groundbreaking study, British geographers used FA to classify British towns. Brian J Berry, at the University of Chicago, and his students made a wide use of the method, applying it to most important cities in the world and exhibiting common social structures. The use of Factor Analysis in Geography, made so easy by modern computers, has been very wide but not always very wise.

Since the vectors extracted are determined by the data matrix, it is not possible to compare factors obtained from different censuses. A solution consists in fusing together several census matrices in a unique table which, then, may be analyzed. This, however, assumes that the definition of the variables has not changed over time and produces very large tables, difficult to manage. A better solution, proposed by psychometricians, groups the data in a « cubic matrix », with three entries (for instance, locations, variables, time periods). A Three-Way Factor Analysis produces then three groups of factors related by a small cubic « core matrix ». This method, which exhibits data evolution over time, has not been widely used in geography. In Los Angeles, however, it has exhibited the role, traditionally ignored, of Downtown as an organizing center for the whole city during several decades.

Spatial Autocorrelation

Spatial autocorrelation statistics measure and analyze the degree of dependency among observations in a geographic space. Classic spatial autocorrelation statistics include Moran's I, Geary's C, Getis's G and the standard deviational ellipse. These statistics require measuring a spatial weights matrix that reflects the intensity of the geographic relationship between observations in a neighborhood, e.g., the distances between neighbors, the lengths of shared border, or whether they fall into a specified directional class such as "west". Classic spatial autocorrelation statistics compare the spatial weights to the covariance relationship at pairs of locations. Spatial autocorrelation that is more positive than expected from random indicate the clustering of similar values across geographic space, while significant negative spatial autocorrelation indicates that neighboring values are more dissimilar than expected by chance, suggesting a spatial pattern similar to a chess board.

Spatial autocorrelation statistics such as Moran's I and Geary's C are global in the sense that they estimate the overall degree of spatial autocorrelation for a dataset. The possibility of spatial heterogeneity suggests that the estimated degree of autocorrelation may vary significantly across geographic space. Local spatial autocorrelation statistics provide estimates disaggregated to the level of the spatial analysis units, allowing assessment of the dependency relationships across space. G statistics compare neighborhoods to a global average and identify local regions of strong autocorrelation. Local versions of the I and C statistics are also available.

Spatial Stratified Heterogeneity

Spatial stratified heterogeneity, referring to the within-strata variance less than the between strata-variance, is ubiquitous in ecological phenomena,such as ecological zones and many ecological variables.Spatial stratified heterogeneity reflects the essence of nature,implies potential distinct mechanisms by strata,suggests possible determinants of the observed process, allows the representativeness of observations of the earth,and enforces the applicability of statistical inferences. Spatial stratified heterogeneity of an attribute can be measured by geographical detector q-statistic:

where a population is partitioned into $h = 1, ..., L$ strata; N stands for the size of the population, σ^2 stands for variance of the attribute. The value of q is within [0, 1], 0 indicates no spatial stratified heterogeneity, 1 indicates perfect spatial stratified heterogeneity. The value of q indicates the percent of the variance of an attribute explained by the stratification.The q follows a noncentral F probability density function.

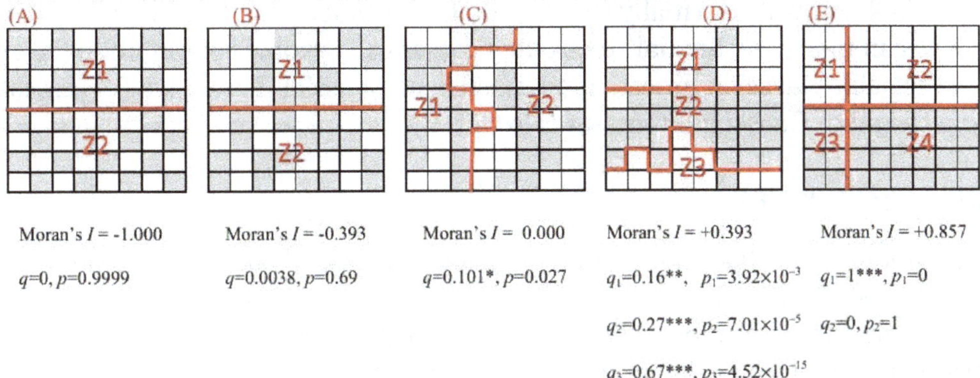

A hand map with different spatial patterns. Note: p is the probability of q-statistic; * denotes statistical significant at level 0.05, ** for 0.001, *** for smaller than 10^{-3};(D) subscripts 1, 2, 3 of q and p denotes the strata Z1+Z2 with Z3,Z1 with Z2+Z3, and Z1 and Z2 and Z3 individually, respectively; (E) subscripts 1 and 2 of q and p denotes the strata Z1+Z2 with Z3+Z4,and Z1+Z3 with Z2+Z4, respectively.

Spatial Interpolation

Spatial interpolation methods estimate the variables at unobserved locations in geographic space based on the values at observed locations. Basic methods include inverse distance weighting: this attenuates the variable with decreasing proximity from the observed location. Kriging is a more sophisticated method that interpolates across space according to a spatial lag relationship that has both systematic and random components. This can accommodate a wide range of spatial relationships for the hidden values between observed locations. Kriging provides optimal estimates given the hypothesized lag relationship, and error estimates can be mapped to determine if spatial patterns exist.

Spatial Regression

Spatial regression methods capture spatial dependency in regression analysis, avoiding statistical problems such as unstable parameters and unreliable significance tests, as well as providing information on spatial relationships among the variables involved. Depending on the specific technique, spatial dependency can enter the regression model as relationships between the independent variables and the dependent, between the dependent variables and a spatial lag of itself, or in the error terms. Geographically weighted regression (GWR) is a local version of spatial regression that generates parameters disaggregated by the spatial units of analysis. This allows assessment of the spatial heterogeneity in the estimated relationships between the independent and dependent variables. The use of Bayesian hierarchical modeling in conjunction with Markov Chain Monte Carlo (MCMC) methods have recently shown to be effective in modeling complex relationships using Poisson-Gamma-CAR, Poisson-lognormal-SAR, or Overdispersed logit models. Spatial stochastic processes, such as Gaussian processes are also increasingly being deployed in spatial regression analysis. Model-based versions of GWR, known as spatially varying coefficient models have been applied to conduct Bayesian inference. Spatial stochastic process can become computationally effective and scalable Gaussian process models, such as Gaussian Predictive Processes and Nearest Neighbor Gaussian Processes (NNGP).

Spatial Interaction

Spatial interaction or "gravity models" estimate the flow of people, material or information between locations in geographic space. Factors can include origin propulsive variables such as the number of commuters in residential areas, destination attractiveness variables such as the amount of office space in employment areas, and proximity relationships between the locations measured in terms such as driving distance or travel time. In addition, the topological, or connective, relationships between areas must be identified, particularly considering the often conflicting relationship between distance and topology; for example, two spatially close neighborhoods may not display any significant interaction if they are separated by a highway. After specifying the functional forms of these relationships, the analyst can estimate model parameters using observed flow data and standard estimation techniques such as ordinary least squares or maximum likelihood. Competing destinations versions of spatial interaction models include the proximity among the destinations (or origins) in addition to the origin-destination proximity; this captures the effects of destination (origin) clustering on flows. Computational methods such as artificial neural networks can also estimate spatial interaction relationships among locations and can handle noisy and qualitative data.

Simulation and Modeling

Spatial interaction models are aggregate and top-down: they specify an overall governing relationship for flow between locations. This characteristic is also shared by urban models such as those based on mathematical programming, flows among economic sectors, or bid-rent theory. An alternative modeling perspective is to represent the system at the highest possible level of disaggregation and study the bottom-up emergence of complex patterns and relationships from behavior and interactions at the individual level.

Complex adaptive systems theory as applied to spatial analysis suggests that simple interactions among proximal entities can lead to intricate, persistent and functional spatial entities at aggregate levels. Two fundamentally spatial simulation methods are cellular automata and agent-based modeling. Cellular automata modeling imposes a fixed spatial framework such as grid cells and specifies rules that dictate the state of a cell based on the states of its neighboring cells. As time progresses, spatial patterns emerge as cells change states based on their neighbors; this alters the conditions for future time periods. For example, cells can represent locations in an urban area and their states can be different types of land use. Patterns that can emerge from the simple interactions of local land uses include office districts and urban sprawl. Agent-based modeling uses software entities (agents) that have purposeful behavior (goals) and can react, interact and modify their environment while seeking their objectives. Unlike the cells in cellular automata, simulysts can allow agents to be mobile with respect to space. For example, one could model traffic flow and dynamics using agents representing individual vehicles that try to minimize travel time between specified origins and destinations. While pursuing minimal travel times, the agents must avoid collisions with other vehicles also seeking to minimize their travel times. Cellular automata and agent-based modeling are complementary modeling strategies. They can be integrated into a common geographic automata system where some agents are fixed while others are mobile.

Multiple-Point Geostatistics (MPS)

Spatial analysis of a conceptual geological model is the main purpose of any MPS algorithm. The method analyzes the spatial statistics of the geological model, called the training image, and generates realizations of the phenomena that honor those input multiple-point statistics.

A recent MPS algorithm used to accomplish this task is the pattern-based method by Honarkhah. In this method, a distance-based approach is employed to analyze the patterns in the training image. This allows the reproduction of the multiple-point statistics, and the complex geometrical features of the training image. Each output of the MPS algorithm is a realization that represents a random field. Together, several realizations may be used to quantify spatial uncertainty.

One of the recent methods is presented by Tahmasebi et al. uses a cross-correlation function to improve the spatial pattern reproduction. They call their MPS simulation method as the CCSIM algorithm. This method is able to quantify the spatial connectivity, variability and uncertainty. Furthermore, the method is not sensitive to any type of data and is able to simulate both categorical and continuous scenarios. CCSIM algorithm is able to be used for any stationary, non-stationary and multivariate systems and it can provide high quality visual appeal model.,

Geographic Information Science and Spatial Analysis

Geographic information systems (GIS) and the underlying geographic information science that advances these technologies have a strong influence on spatial analysis. The increasing ability to capture and handle geographic data means that spatial analysis is occurring within increasingly data-rich environments. Geographic data capture systems include remotely sensed imagery, environmental monitoring systems such as intelligent transportation systems, and location-aware technologies such as mobile devices that can report location in near-real time. GIS provide platforms for managing these data, computing spatial relationships such as distance, connectivity and directional relationships between spatial units, and visualizing both the raw data and spatial analytic results within a cartographic context.

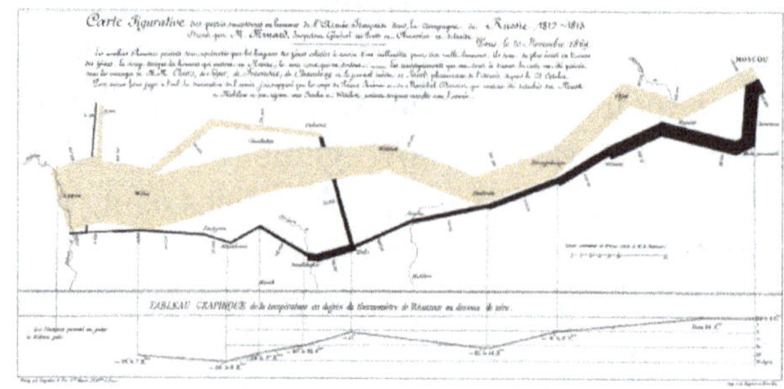

This flow map of Napoleon's ill-fated march on Moscow is an early and celebrated example of geovisualization. It shows the army's direction as it traveled, the places the troops passed through, the size of the army as troops died from hunger and wounds, and the freezing temperatures they experienced.

Content

- Spatial location: Transfer positioning information of space objects with the help of space coordinate system. Projection transformation theory is the foundation of spatial object representation.
- Spatial distribution: the similar spatial object groups positioning information, including distribution, trends, contrast etc..
- Spatial form: the geometric shape of the spatial objects
- Spatial space: the space objects' approaching degree
- Spatial relationship: relationship between spatial objects, including topological, orientation, similarity, etc..

Geovisualization (GVis) combines scientific visualization with digital cartography to support the exploration and analysis of geographic data and information, including the results of spatial analysis or simulation. GVis leverages the human orientation towards visual information processing in the exploration, analysis and communication of geographic data and information. In contrast with traditional cartography, GVis is typically three- or four-dimensional (the latter including time) and user-interactive.

Geographic knowledge discovery (GKD) is the human-centered process of applying efficient computational tools for exploring massive spatial databases. GKD includes geographic data mining, but also encompasses related activities such as data selection, data cleaning and pre-processing, and interpretation of results. GVis can also serve a central role in the GKD process. GKD is based on the premise that massive databases contain interesting (valid, novel, useful and understandable) patterns that standard analytical techniques cannot find. GKD can serve as a hypothesis-generating process for spatial analysis, producing tentative patterns and relationships that should be confirmed using spatial analytical techniques.

Spatial Decision Support Systems (SDSS) take existing spatial data and use a variety of mathematical models to make projections into the future. This allows urban and regional planners to test intervention decisions prior to implementation.

Geotagged Photograph

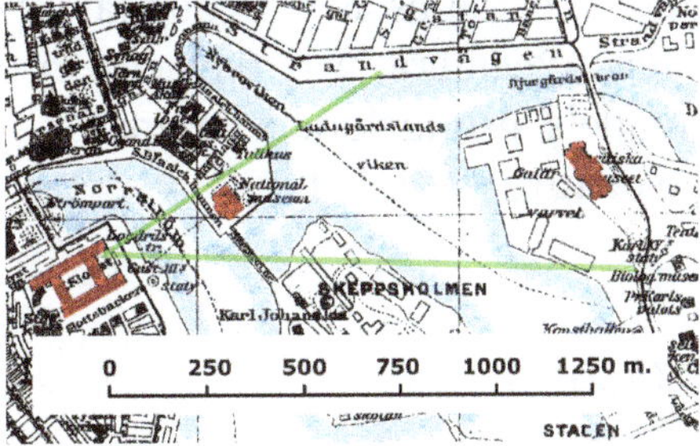

Above: Map showing photographer's view (59°19′39″N 18°04′21″E59.3275°N 18.0725°E) of two buildings at a distance of 270 meters (59°19′42″N 18°04′38″E59.3284°N 18.0772°E) and 1200 meters (59°19′45″N 18°05′35″E59.3291°N 18.0931°E). Below: Which of the three locations should be associated with the resulting photo?

A geotagged photograph is a photograph which is associated with a geographical location by geotagging. Usually this is done by assigning at least a latitude and longitude to the image, and optionally altitude, compass bearing and other fields may also be included.

In theory, every part of a picture can be tied to a geographic location, but in the most typical application, only the position of the photographer is associated with the entire digital image. This has implications for search and retrieval. For example, photos of a mountain summit can be taken from different positions miles apart. To find all images of a particular summit in an image database, all photos taken within a reasonable distance must be considered. The point position of the photographer can in some cases include the bearing, the direction the camera was pointing, as well as the elevation and the DOP.

Methods of Geotagging Photographs

There are a few methods of geotagging photographs, either automatic or manual. Automatic methods provide the easiest and most precise method of geotagging an image, providing that a good signal has been acquired at the time of taking the photo.

Automatic using a Built-in GPS

Several manufacturers offer cameras with a built-in GPS receiver, but most cameras with this capability are camera phones as camera manufacturers after initial experience in the market came to treat GPS cameras as a niche market. The 2008 Nikon P6000, for example, an early geotagging camera, was replaced in 2010 by the P7000 which lacked that feature. Some models also include a compass to indicate the direction the camera was facing when the picture was taken.

- Canon EOS 6D
- Canon Powershot SX280HS
- Panasonic Lumix DMC-TZ10
- Sony Alpha 55V (DSLR)
- Sony Alpha 65V (DSLR)
- Nikon Coolpix P6000
- Some mobile phones with assisted GPS use the cell phone network to speed GPS acquisition times.

Automatic using a Connected GPS

The D1X and D1H that Nikon introduced in 2002 included a GPS interface. In 2006 the first special GPS receiver for Nikon was produced by Dawntech. Since 2009 Nikon has sold its own Geotagger GP-1. Canon uses the USB socket on the wireless file transmitter unit (WFT) as the GPS interface.

Geotagger "Solmeta N2 Compass" + Nikon D5000

Some digital cameras and camera phones support an external GPS receiver connected by cable, or inserted into the memory card slot or flash shoe. The Samsung SH100 can connect using Wi-Fi to get position data from a GPS-enabled smartphone. Generally the relevant GPS data is automatically stored in the photo's Exif information when the photo is taken. A connected GPS will generally remain switched on continuously, requiring power, and will then have location information available immediately when the camera is switched on.

Many GPS-ready cameras are currently available, made by manufacturers such as Nikon, Fujifilm, Sony and Panasonic. Automatic geotagging combined with real-time transfer and publishing results in real-time geotagging.

Synchronizing with A Separate GPS

Most cameras sold today do not contain a built-in GPS receiver; however, an external location-aware device, such as a hand-held GPS logger, can still be used with a non-GPS digital camera for geotagging. The photo is taken without geographical information and is processed later using software in conjunction with the GPS data. Timestamps made by the camera can be compared with timestamps in the recorded GPS information, provided that the clocks in the separate devices can be synchronized. The resulting coordinates can then be added to the Exif information of the photo.

Manual Geotagging

Location information can also be added to photos, for example via its Exif specification that has fields for longitude/latitude, even if no GPS device was present when the photo was taken.

The information can be entered by directly giving the coordinates or by selecting a location from a map using software tools. Some tools allow entry of tags such as city, postal code or a street address. Geocoding and reverse geocoding can be used to convert between locations and addresses.

Manual geotagging also introduces possibilities of error, where a photograph's location is incorrectly represented by wrong coordinates. An advanced comparative analysis of such photos with the total collection set of all photos available from the surrounding coordinates, needs to be done to single out and flag such photos, but such a software's value, need and purpose could be limited in today's environment where almost every smartphone and camera have geotagging built-in and users do not need to manually enter this information.

Remote Standoff Capture

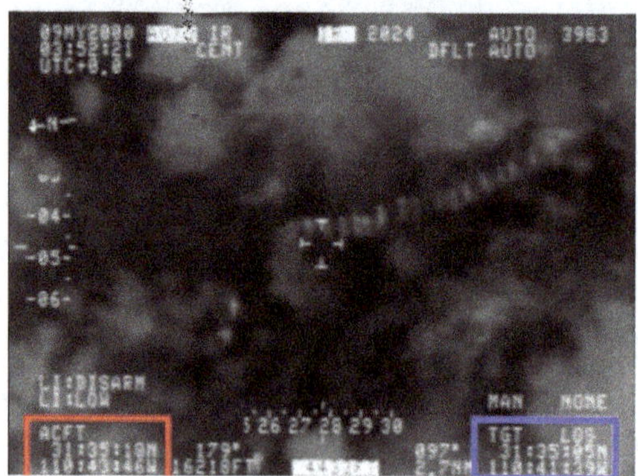

Screenshot from a U.S. Customs and Border Protection Predator UAV, showing the GPS position of the aircraft (red) and the target (blue)

Some manufacturers of military and professional mapping-grade GPS instruments have integrated a GPS receiver with a laser rangefinder and digital camera. These multi-functional tools are able to determine a remote subject's GPS position by calculating the subject's geographic location relative to the camera's GPS position. These instruments are commonly used in military applications when an aircraft or operator is targeting an area, the position is inaccessible (for example over a valley or wetland), there are personal health & safety concerns (motorway traffic), or the user wants to quickly capture multiple targets from a single, safe position (trees, street signage and furniture).

Civilian integrated GPS cameras with rangefinders and remote standoff capability are currently available made by manufacturers such as Ricoh and Surveylab.

Uses

When geotagged photos are uploaded to online sharing communities such as Flickr, Panoramio or Moblog, the photo can be placed onto a map to view the location the photo was taken. In this way, users can browse photos from a map, search for photos from a given area, and find related photos of the same place from other users.

Many smartphones automatically geotag their photos by default. Photographers who prefer not to reveal their location can turn this feature off. Additionally smartphones can use their GPS to geotag photos taken with an external camera.

Geotagged photo location stamped with GPStamper

Geotagged photos may be visually stamped with their GPS location information using software tools. A stamped photo affords universal and cross-platform viewing of the photo's location, and offers the security of retaining that location information in the event of metadata corruption, or if file metadata is stripped from a photo, e.g. when uploading to various online photo sharing communities.

Geotagging is also being used to determine social patterns. For example, Now app uses geotagged Instagram photos to find nearby events happening now.

Remote Sensing

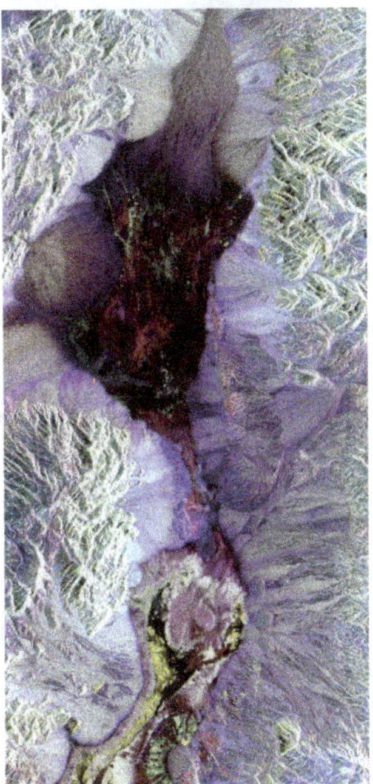

Synthetic aperture radar image of Death Valley colored using polarimetry.

Remote sensing is the acquisition of information about an object or phenomenon without making physical contact with the object and thus in contrast to on-site observation. Remote sensing is used in numerous fields, including geography and most Earth Science disciplines (for example, hydrology, ecology, oceanography, glaciology, geology); it also has military, intelligence, commercial, economic, planning, and humanitarian applications.

In current usage, the term “remote sensing” generally refers to the use of satellite- or aircraft-based sensor technologies to detect and classify objects on Earth, including on the surface and in the atmosphere and oceans, based on propagated signals (e.g. electromagnetic radiation). It may be split into “active” remote sensing (i.e., when a signal is emitted by a satellite or aircraft and its reflection by the object is detected by the sensor) and “passive” remote sensing (i.e., when the reflection of sunlight is detected by the sensor).

Overview

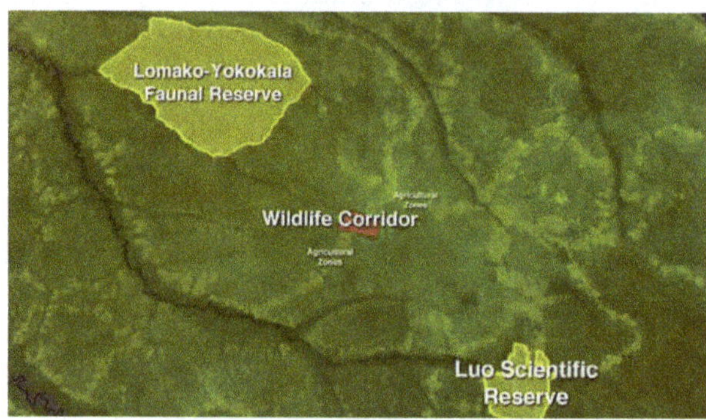

This video is about how Landsat was used to identify areas of conservation in the Democratic Republic of the Congo, and how it was used to help map an area called MLW in the north

Passive sensors gather radiation that is emitted or reflected by the object or surrounding areas. Reflected sunlight is the most common source of radiation measured by passive sensors. Examples of passive remote sensors include film photography, infrared, charge-coupled devices, and radiometers. Active collection, on the other hand, emits energy in order to scan objects and areas whereupon a sensor then detects and measures the radiation that is reflected or backscattered from the target. RADAR and LiDAR are examples of active remote sensing where the time delay between emission and return is measured, establishing the location, speed and direction of an object.

Remote sensing makes it possible to collect data of dangerous or inaccessible areas. Remote sensing applications include monitoring deforestation in areas such as the Amazon Basin, glacial features in Arctic and Antarctic regions, and depth sounding of coastal and ocean depths. Military collection during the Cold War made use of stand-off collection of data about dangerous border areas. Remote sensing also replaces costly and slow data collection on the ground, ensuring in the process that areas or objects are not disturbed.

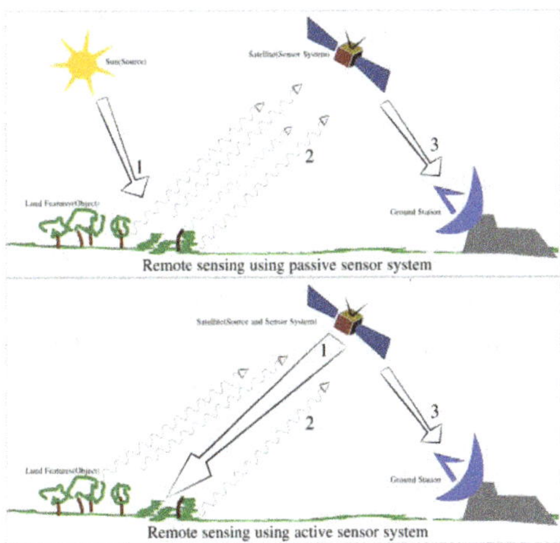

Illustration of Remote Sensing

Orbital platforms collect and transmit data from different parts of the electromagnetic spectrum, which in conjunction with larger scale aerial or ground-based sensing and analysis, provides researchers with enough information to monitor trends such as El Niño and other natural long and short term phenomena. Other uses include different areas of the earth sciences such as natural resource management, agricultural fields such as land usage and conservation, and national security and overhead, ground-based and stand-off collection on border areas.

Data Acquisition Techniques

The basis for multispectral collection and analysis is that of examined areas or objects that reflect or emit radiation that stand out from surrounding areas. For a summary of major remote sensing satellite systems see the overview table.

Applications of Remote Sensing

- Conventional radar is mostly associated with aerial traffic control, early warning, and certain large scale meteorological data. Doppler radar is used by local law enforcements' monitoring of speed limits and in enhanced meteorological collection such as wind speed and direction within weather systems in addition to precipitation location and intensity. Other types of active collection includes plasmas in the ionosphere. Interferometric synthetic aperture radar is used to produce precise digital elevation models of large scale terrain.

- Laser and radar altimeters on satellites have provided a wide range of data. By measuring the bulges of water caused by gravity, they map features on the seafloor to a resolution of a mile or so. By measuring the height and wavelength of ocean waves, the altimeters measure wind speeds and direction, and surface ocean currents and directions.

- Ultrasound (acoustic) and radar tide gauges measure sea level, tides and wave direction in coastal and offshore tide gauges.
- Light detection and ranging (LIDAR) is well known in examples of weapon ranging, laser illuminated homing of projectiles. LIDAR is used to detect and measure the concentration of various chemicals in the atmosphere, while airborne LIDAR can be used to measure heights of objects and features on the ground more accurately than with radar technology. Vegetation remote sensing is a principal application of LIDAR.
- Radiometers and photometers are the most common instrument in use, collecting reflected and emitted radiation in a wide range of frequencies. The most common are visible and infrared sensors, followed by microwave, gamma ray and rarely, ultraviolet. They may also be used to detect the emission spectra of various chemicals, providing data on chemical concentrations in the atmosphere.
- Stereographic pairs of aerial photographs have often been used to make topographic maps by imagery and terrain analysts in trafficability and highway departments for potential routes, in addition to modelling terrestrial habitat features.
- Simultaneous multi-spectral platforms such as Landsat have been in use since the 70's. These thematic mappers take images in multiple wavelengths of electro-magnetic radiation (multi-spectral) and are usually found on Earth observation satellites, including (for example) the Landsat program or the IKONOS satellite. Maps of land cover and land use from thematic mapping can be used to prospect for minerals, detect or monitor land usage, detect invasive vegetation, deforestation, and examine the health of indigenous plants and crops, including entire farming regions or forests. Landsat images are used by regulatory agencies such as KYDOW to indicate water quality parameters including Secchi depth, chlorophyll a density and total phosphorus content. Weather satellites are used in meteorology and climatology.
- Hyperspectral imaging produces an image where each pixel has full spectral information with imaging narrow spectral bands over a contiguous spectral range. Hyperspectral imagers are used in various applications including mineralogy, biology, defence, and environmental measurements.
- Within the scope of the combat against desertification, remote sensing allows to follow up and monitor risk areas in the long term, to determine desertification factors, to support decision-makers in defining relevant measures of environmental management, and to assess their impacts.

Geodetic

- Geodetic remote sensing can be gravimetric or geometric. Overhead gravity data collection was first used in aerial submarine detection. This data revealed minute perturbations in the Earth's gravitational field that may be used to determine changes in the mass distribution of the Earth, which in turn may be used for geophysical studies, as in GRACE (satellite). Geometric remote sensing includes position and deformation imaging using InSAR, lidar, etc.

Acoustic and Near-Acoustic

- Sonar: *passive sonar*, listening for the sound made by another object (a vessel, a whale etc.); *active sonar*, emitting pulses of sounds and listening for echoes, used for detecting, ranging and measurements of underwater objects and terrain.
- Seismograms taken at different locations can locate and measure earthquakes (after they occur) by comparing the relative intensity and precise timings.
- Ultrasound: Ultrasound sensors, that emit high frequency pulses and listening for echoes, used for detecting water waves and water level, as in tide gauges or for towing tanks.

To coordinate a series of large-scale observations, most sensing systems depend on the following: platform location and the orientation of the sensor. High-end instruments now often use positional information from satellite navigation systems. The rotation and orientation is often provided within a degree or two with electronic compasses. Compasses can measure not just azimuth (i. e. degrees to magnetic north), but also altitude (degrees above the horizon), since the magnetic field curves into the Earth at different angles at different latitudes. More exact orientations require gyroscopic-aided orientation, periodically realigned by different methods including navigation from stars or known benchmarks.

Data Processing

Generally speaking, remote sensing works on the principle of the *inverse problem*. While the object or phenomenon of interest (the state) may not be directly measured, there exists some other variable that can be detected and measured (the observation) which may be related to the object of interest through a calculation. The common analogy given to describe this is trying to determine the type of animal from its footprints. For example, while it is impossible to directly measure temperatures in the upper atmosphere, it is possible to measure the spectral emissions from a known chemical species (such as carbon dioxide) in that region. The frequency of the emissions may then be related via thermodynamics to the temperature in that region.

The quality of remote sensing data consists of its spatial, spectral, radiometric and temporal resolutions.

Spatial resolution

The size of a pixel that is recorded in a raster image – typically pixels may correspond to square areas ranging in side length from 1 to 1,000 metres (3.3 to 3,280.8 ft).

Spectral resolution

The wavelength width of the different frequency bands recorded – usually, this is related to the number of frequency bands recorded by the platform. Current Landsat collection is that of seven bands, including several in the infra-red spectrum, ranging from a spectral resolution of 0.07 to 2.1 μm. The Hyperion sensor on Earth Observing-1 resolves 220 bands from 0.4 to 2.5 μm, with a spectral resolution of 0.10 to 0.11 μm per band.

Radiometric resolution

The number of different intensities of radiation the sensor is able to distinguish. Typically, this ranges from 8 to 14 bits, corresponding to 256 levels of the gray scale and up to 16,384 intensities or "shades" of colour, in each band. It also depends on the instrument noise.

Temporal resolution

The frequency of flyovers by the satellite or plane, and is only relevant in time-series studies or those requiring an averaged or mosaic image as in deforesting monitoring. This was first used by the intelligence community where repeated coverage revealed changes in infrastructure, the deployment of units or the modification/introduction of equipment. Cloud cover over a given area or object makes it necessary to repeat the collection of said location.

In order to create sensor-based maps, most remote sensing systems expect to extrapolate sensor data in relation to a reference point including distances between known points on the ground. This depends on the type of sensor used. For example, in conventional photographs, distances are accurate in the center of the image, with the distortion of measurements increasing the farther you get from the center. Another factor is that of the platen against which the film is pressed can cause severe errors when photographs are used to measure ground distances. The step in which this problem is resolved is called georeferencing, and involves computer-aided matching of points in the image (typically 30 or more points per image) which is extrapolated with the use of an established benchmark, "warping" the image to produce accurate spatial data. As of the early 1990s, most satellite images are sold fully georeferenced.

In addition, images may need to be radiometrically and atmospherically corrected.

Radiometric correction

Allows to avoid radiometric errors and distortions. The illumination of objects on the Earth surface is uneven because of different properties of the relief. This factor is taken into account in the method of radiometric distortion correction. Radiometric correction gives a scale to the pixel values, e. g. the monochromatic scale of 0 to 255 will be converted to actual radiance values.

Topographic correction (also called terrain correction)

In rugged mountains, as a result of terrain, the effective illumination of pixels varies considerably. In a remote sensing image, the pixel on the shady slope receives weak illumination and has a low radiance value, in contrast, the pixel on the sunny slope receives strong illumination and has a high radiance value. For the same object, the pixel radiance value on the shady slope will be different from that on the sunny slope. Additionally, different objects may have similar radiance values. These ambiguities seriously affected remote sensing image information extraction accuracy in mountainous areas. It became the main obstacle to further application of remote sensing images. The purpose of topographic correction is to eliminate this effect, recovering the true reflectivity or radiance of objects in horizontal conditions. It is the premise of quantitative remote sensing application.

Atmospheric correction

Elimination of atmospheric haze by rescaling each frequency band so that its minimum value (usually realised in water bodies) corresponds to a pixel value of 0. The digitizing of data also makes it possible to manipulate the data by changing gray-scale values.

Interpretation is the critical process of making sense of the data. The first application was that of aerial photographic collection which used the following process; spatial measurement through the use of a light table in both conventional single or stereographic coverage, added skills such as the use of photogrammetry, the use of photomosaics, repeat coverage, Making use of objects' known dimensions in order to detect modifications. Image Analysis is the recently developed automated computer-aided application which is in increasing use.

Object-Based Image Analysis (OBIA) is a sub-discipline of GIScience devoted to partitioning remote sensing (RS) imagery into meaningful image-objects, and assessing their characteristics through spatial, spectral and temporal scale.

Old data from remote sensing is often valuable because it may provide the only long-term data for a large extent of geography. At the same time, the data is often complex to interpret, and bulky to store. Modern systems tend to store the data digitally, often with lossless compression. The difficulty with this approach is that the data is fragile, the format may be archaic, and the data may be easy to falsify. One of the best systems for archiving data series is as computer-generated machine-readable ultrafiche, usually in typefonts such as OCR-B, or as digitized half-tone images. Ultrafiches survive well in standard libraries, with lifetimes of several centuries. They can be created, copied, filed and retrieved by automated systems. They are about as compact as archival magnetic media, and yet can be read by human beings with minimal, standardized equipment.

Data Processing Levels

To facilitate the discussion of data processing in practice, several processing "levels" were first defined in 1986 by NASA as part of its Earth Observing System and steadily adopted since then, both internally at NASA (e. g.,) and elsewhere (e. g.,); these definitions are:

Level	Description
0	Reconstructed, unprocessed instrument and payload data at full resolution, with any and all communications artifacts (e. g., synchronization frames, communications headers, duplicate data) removed.
1a	Reconstructed, unprocessed instrument data at full resolution, time-referenced, and annotated with ancillary information, including radiometric and geometric calibration coefficients and georeferencing parameters (e. g., platform ephemeris) computed and appended but not applied to the Level 0 data (or if applied, in a manner that level 0 is fully recoverable from level 1a data).
1b	Level 1a data that have been processed to sensor units (e. g., radar backscatter cross section, brightness temperature, etc.); not all instruments have Level 1b data; level 0 data is not recoverable from level 1b data.
2	Derived geophysical variables (e. g., ocean wave height, soil moisture, ice concentration) at the same resolution and location as Level 1 source data.
3	Variables mapped on uniform spacetime grid scales, usually with some completeness and consistency (e. g., missing points interpolated, complete regions mosaicked together from multiple orbits, etc.).
4	Model output or results from analyses of lower level data (i. e., variables that were not measured by the instruments but instead are derived from these measurements).

A Level 1 data record is the most fundamental (i. e., highest reversible level) data record that has significant scientific utility, and is the foundation upon which all subsequent data sets are produced. Level 2 is the first level that is directly usable for most scientific applications; its value is much greater than the lower levels. Level 2 data sets tend to be less voluminous than Level 1 data because they have been reduced temporally, spatially, or spectrally. Level 3 data sets are generally smaller than lower level data sets and thus can be dealt with without incurring a great deal of data handling overhead. These data tend to be generally more useful for many applications. The regular spatial and temporal organization of Level 3 datasets makes it feasible to readily combine data from different sources.

While these processing levels are particularly suitable for typical satellite data processing pipelines, other data level vocabularies have been defined and may be appropriate for more heterogeneous workflows.

History

The TR-1 reconnaissance/surveillance aircraft.

The *2001 Mars Odyssey* used spectrometers and imagers to hunt for evidence of past or present water and volcanic activity on Mars.

The modern discipline of remote sensing arose with the development of flight. The balloonist G. Tournachon (alias Nadar) made photographs of Paris from his balloon in 1858. Messenger pigeons, kites, rockets and unmanned balloons were also used for early images. With the exception of balloons, these first, individual images were not particularly useful for map making or for scientific purposes.

Systematic aerial photography was developed for military surveillance and reconnaissance purposes beginning in World War I and reaching a climax during the Cold War with the use of modified combat aircraft such as the P-51, P-38, RB-66 and the F-4C, or specifically designed collection platforms such as the U2/TR-1, SR-71, A-5 and the OV-1 series both in overhead and stand-off collection. A more recent development is that of increasingly smaller sensor pods such as those used by law enforcement and the military, in both manned and unmanned platforms. The advantage of this approach is that this requires minimal modification to a given airframe. Later imaging technologies would include Infra-red, conventional, Doppler and synthetic aperture radar.

The development of artificial satellites in the latter half of the 20th century allowed remote sensing to progress to a global scale as of the end of the Cold War. Instrumentation aboard various Earth observing and weather satellites such as Landsat, the Nimbus and more recent missions such as RADARSAT and UARS provided global measurements of various data for civil, research, and military purposes. Space probes to other planets have also provided the opportunity to conduct remote sensing studies in extraterrestrial environments, synthetic aperture radar aboard the Magellan spacecraft provided detailed topographic maps of Venus, while instruments aboard SOHO allowed studies to be performed on the Sun and the solar wind, just to name a few examples.

Recent developments include, beginning in the 1960s and 1970s with the development of image processing of satellite imagery. Several research groups in Silicon Valley including NASA Ames Research Center, GTE, and ESL Inc. developed Fourier transform techniques leading to the first notable enhancement of imagery data. In 1999 the first commercial satellite (IKONOS) collecting very high resolution imagery was launched.

Training and Education

At most universities remote sensing is associated with Geography departments. Remote Sensing has a growing relevance in the modern information society. It represents a key technology as part of the aerospace industry and bears increasing economic relevance – new sensors e.g. TerraSAR-X and RapidEye are developed constantly and the demand for skilled labour is increasing steadily. Furthermore, remote sensing exceedingly influences everyday life, ranging from weather forecasts to reports on climate change or natural disasters. As an example, 80% of the German students use the services of Google Earth; in 2006 alone the software was downloaded 100 million times. But studies have shown that only a fraction of them know more about the data they are working with. There exists a huge knowledge gap between the application and the understanding of satellite images. Remote sensing only plays a tangential role in schools, regardless of the political claims to strengthen the support for teaching on the subject. A lot of the computer software explicitly developed for school lessons has not yet been implemented due to its complexity. Thereby, the subject is either not at all integrated into the curriculum or does not pass the step of an interpretation of analogue images. In fact, the subject of remote sensing requires a consolidation of physics and mathematics as well as competences in the fields of media and methods apart from the mere visual interpretation of satellite images.

Many teachers have great interest in the subject "remote sensing", being motivated to integrate this topic into teaching, provided that the curriculum is considered. In many cases, this encouragement fails because of confusing information. In order to integrate remote sensing in a sustainable manner organizations like the EGU or digital earth encourages the development of learning modules and learning portals (e.g. FIS – Remote Sensing in School Lessons or Landmap – Spatial Discovery) promoting media and method qualifications as well as independent working.

Remote Sensing Software

Remote sensing data are processed and analyzed with computer software, known as a remote sensing application. A large number of proprietary and open source applications exist to process remote sensing data. Remote sensing software packages include:

- ERDAS IMAGINE from Hexagon Geospatial (Separated from Intergraph SG&I),
- PCI Geomatica made by PCI Geomatics,
- TacitView from 2d3
- SOCET GXP from BAE Systems,
- TNTmips from MicroImages,
- IDRISI from Clark Labs,
- eCognition from Trimble,
- and RemoteView made by Overwatch Textron Systems.
- Dragon/ips is one of the oldest remote sensing packages still available, and is in some cases free.
- ENVI/IDL from Exelis Visual Information Solutions,

Open source remote sensing software includes:

- Opticks (software),
- Orfeo toolbox
- Others mixing remote sensing and GIS capabilities are: GRASS GIS, ILWIS, QGIS, and TerraLook.

According to an NOAA Sponsored Research by Global Marketing Insights, Inc. the most used applications among Asian academic groups involved in remote sensing are as follows: ERDAS 36% (ERDAS IMAGINE 25% & ERMapper 11%); ESRI 30%; ITT Visual Information Solutions ENVI 17%; MapInfo 17%.

Among Western Academic respondents as follows: ESRI 39%, ERDAS IMAGINE 27%, MapInfo 9%, and AutoDesk 7%.

Rubbersheeting

In cartography, rubbersheeting refers to the process by which a layer is distorted to allow it to be seamlessly joined to an adjacent geographic layer of matching imagery, such as satellite imagery (most commonly vector cartographic data) which are digital maps. This is sometimes referred to as image-to-vector conflation. Often this has to be done when layers created from adjacent map sheets are joined together. Rubber-sheeting is necessary because the imagery and the vector data will rarely match up correctly due to various reasons, such as the angle at which the image was taken, the curvature of the surface of the earth, minor movements in the imaging platform (such as a satellite or aircraft), and other errors in the imagery.

Applications in History and Historical Geography

Rubbersheeting is a useful technique in HGIS, where it is used to digitize and add old maps as feature layers in a modern GIS. Before aerial photography arrived, most maps were highly inaccurate by modern standards. Rubbersheeting may improve the value of such sources and make them easier to compare to modern maps.

Software

- ESRI's ArcGIS 8.3+ has the capability of rubbersheeting vector data, and ArcMap 9.2+ may also rubber-sheet raster layers.
- Autodesk's AutoCAD Map 3D and AutoCAD Civil 3D (which includes most of AutoCAD Map 3D's functionality) allows a user to rubbersheet vector data, and Autodesk's Raster Design (an add-in product for AutoCAD-based products) allows a user to rubbersheet raster data.
- Blue Marble Geographics' Global Mapper allows a user to rubbersheet raster data.
- Cadcorp Spatial Information System software (SIS Map Modeller) is offering a tool for rubbersheeting data layers.
- QGIS Georeferencer plug-in provides a number of transformation types including Thin Plate Spline, which enables full rubber-sheeting. QGIS is a free open-source GIS package.

References

- Schowengerdt, Robert A. (2007). Remote sensing: models and methods for image processing (3rd ed.). Academic Press. p. 2. ISBN 978-0-12-369407-2.
- Schott, John Robert (2007). Remote sensing: the image chain approach (2nd ed.). Oxford University Press. p. 1. ISBN 978-0-19-517817-3.
- Liu, Jian Guo & Mason, Philippa J. (2009). Essential Image Processing for GIS and Remote Sensing. Wiley-Blackwell. p. 4. ISBN 978-0-470-51032-2.
- Wang, JF; Zhang, TL; Fu, BJ (2016). "A measure of spatial stratified heterogeneity". Ecological Indicators. 67: 250–256. doi:10.1016/j.ecolind.2016.02.052.

4

Geocoding: An Overview

Geocoding is the practice of converting a location to an address on the Earth's surface. The opposite of geocoding is reverse geocoding; where the process is of back coding a location to a readable address. This chapter is an overview of the subject matter incorporating all the major aspects of geocoding.

Geocoding

Geocoding is the computational process of transforming a (postal) address description to a location on the Earth's surface (spatial representation in numerical coordinates). Reverse geocoding, on the other hand, converts the inputted geographic coordinates to a description of a location, usually the name of a place or a postal address. Geocoding relies on a computer representation of the street network. Geocoding is sometimes used for conversion from ZIP codes or postal codes to coordinates, occasionally for the conversion of parcel identifiers to centroid coordinates.

Geocoding (*verb*): The act of transforming an address text into a valid spatial representation.

Geocoder (*noun*): A piece of software or a (web) service that implements a geocoding process i.e. a set of interrelated components in the form of operations, algorithms, and data sources that work together to produce a spatial representation for descriptive locational references.

Geocode (*noun*): A spatial representation of a descriptive locational reference.

The geographic coordinates representing locations often vary greatly in positional accuracy. Examples include building centroids, land parcels, street addresses, postal code centroids (e.g. ZIP codes, CEDEX), and Administrative Boundary Centroids.

History

Geocoding — a subset of Geographic Information System (GIS) spatial analysis — has been a subject of interest since the early 1960s.

1960s

In 1960, the first operational GIS — named the Canada Geographic Information System

(CGIS) — was invented by Dr. Roger Tomlinson, who has since been acknowledged as the father of GIS. The CGIS was used to store and analyze data collected for the Canada Land Inventory, which mapped information about agriculture, wildlife, and forestry at a scale of 1:50,000, in order to regulate land capability for rural Canada. However, the CGIS lasted until the 1990s and was never available commercially.

On July 1, 1963, five-digit ZIP codes were introduced nationwide by the United States Post Office Department (USPOD). In 1983, nine-digit ZIP+4 codes were brought about as an extra identifier in more accurately locating addresses.

In 1964, the Harvard Laboratory for Computer Graphics and Spatial Analysis developed groundbreaking software code — e.g. GRID, and SYMAP — all of which were sources for commercial development of GIS.

In 1967, a team at the Census Bureau — including the mathematician James Corbett and Donald Cooke — invented Dual Independent Map Encoding (DIME) — the first modern vector mapping model — which ciphered address ranges into street network files and incorporated the "percent along" geocoding algorithm. Still in use by platforms such as Google Maps and MapQuest, the "percent along" algorithm denotes where a matched address is located along a reference feature as a percentage of the reference feature's total length. DIME was intended for the use of the United States Census Bureau, and it involved accurately mapping block faces, digitizing nodes representing street intersections, and forming spatial relationships. New Haven, Connecticut was the first city on Earth with a geocodable streets network database.

1980s

In the late 1970s, two main public domain geocoding platforms were in development: GRASS GIS and MOSS. The early 1980s saw the rise of many more commercial vendors of geocoding software, namely Intergraph, ESRI, CARIS, ERDAS, and MapInfo Corporation. These platforms merged the 1960s approach of separating spatial information with the approach of organizing this spatial information into database structures.

In 1986, Mapping Display and Analysis System (MIDAS) became the first desktop geocoding software, designed for the DOS operating system. Geocoding was elevated from the research department into the business world with the acquisition of MIDAS by MapInfo. MapInfo has since been acquired by Pitney Bowes, and has pioneered in merging geocoding with business intelligence; allowing location intelligence to provide solutions for the public and private sectors.

1990s

The end of the 20th century had seen geocoding become more user-oriented, especially via open-source GIS software. Mapping applications and geospatial data had become more accessible over the Internet.

Because the mail-out/mail-back technique was so successful in the 1980 Census, the U.S. Bureau of Census was able to put together a large geospatial database, using interpolated street geocoding. This database – along with the Census' nationwide coverage of households – allowed for the birth of TIGER (Topologically Integrated Geographic Encoding and Referencing).

Containing address ranges instead of individual addresses, TIGER has since been implemented in nearly all geocoding software platforms used today. By the end of the 1990 Census, TIGER "contained a latitude/longitude-coordinate for more than 30 million feature intersections and endpoints and nearly 145 million feature 'shape' points that defined the more than 42 million feature segments that outlined more than 12 million polygons."

TIGER was the breakthrough for "big data" geospatial solutions.

2000s

The early 2000s saw the rise of Coding Accuracy Support System (CASS) address standardization. The CASS certification is offered to all software vendors and advertising mailers who want the United States Postal Services (USPS) to assess the quality of their address-standardizing software. The annually renewed CASS certification is based on delivery point codes, ZIP codes, and ZIP+4 codes. Adoption of a CASS certified software by software vendors allows them to receive discounts in bulk mailing and shipping costs. They can benefit from increased accuracy and efficiency in those bulk mailings, after having a certified database. In the early 2000s, geocoding platforms were also able to support multiple datasets.

In 2003, geocoding platforms were capable of merging postal codes with street data, updated monthly. This process became known as "conflation".

Beginning in 2005, geocoding platforms included parcel-centroid geocoding. Parcel-centroid geocoding allowed for a lot of precision in geocoding an address. For example, parcel-centroid allowed a geocoder to determine the centroid of a specific building or lot of land. Platforms were now also able to determine the elevation of specific parcels.

2005 also saw the introduction of the Assessor's Parcel Number (APN). A jurisdiction's tax assessor was able to assign this number to parcels of real estate. This allowed for proper identification and record-keeping. An APN is important for geocoding an area which is covered by a gas or oil lease, and indexing property tax information provided to the public.

In 2006, Reverse Geocoding and reverse APN lookup were introduced to geocoding platforms. This involved geocoding a numerical point location – with a longitude and latitude – to a textual, readable address.

2008 and 2009 saw the growth of interactive, user-oriented geocoding platforms – namely MapQuest, Google Maps, Bing Maps, and Global Positioning Systems (GPS). These platforms were made even more accessible to the public with the simultaneous growth of the mobile industry, specifically smartphones.

2010s

This current decade has seen vendors fully supporting geocoding and reverse geocoding globally. Cloud-based geocoding application programming interface (API) and on-premise geocoding has allowed for a greater match rate, greater precision, and greater speed. There is now a popularity in the idea of geocoding being able to influence business decisions. This is the integration between the geocoding process and business intelligence.

The future of geocoding also involves three-dimensional geocoding, indoor geocoding, and multiple language returns for the geocoding platforms.

Geocoding Process

Geocoding is a task which involves multiple datasets and processes, all of which work together. A geocoder is made of two important components: a reference dataset and the geocoding algorithm. Each of these components are made up of sub-operations and sub-components. Without understanding how these geocoding processes work, it is difficult to make informed business decisions based on geocoding.

Input Data

Input data are the descriptive, textual information (address or building name) which the user wants to turn into numerical, spatial data (latitude and longitude) – through the process of geocoding.

Classification of Input Data

Input data is classified into two categories: relative input data and absolute input data.

Relative Input Data

Relative input data are the textual descriptions of a location which, alone, cannot output a spatial representation of that location. Such data outputs a relative geocode, which is dependent and geographically relative of other reference locations. An example of a relative geocode is address-interpolation using areal units or line vectors. "Across the street from the Empire State Building" is an example of a relative input data. The location being sought cannot be determined without identifying the Empire State Building. Geocoding platforms often do not support such relative locations, but advances are being made in this direction.

Absolute Input Data

Absolute input data are the textual descriptions of a location which, alone, can output a spatial representation of that location. This data type outputs an absolute known location independently of other locations. For example, USPS ZIP codes; USPS ZIP+4 codes; complete and partial postal addresses; USPS PO boxes; rural routes; cities; counties; intersections; and named places can all be referenced in a data source absolutely.

When there is a lot of variability in the way addresses can be represented – such as too much input data or too little input data – geocoders use address normalization and address standardization in order to resolve this problem.

Processing of Input Data

Address Interpolation

A simple method of geocoding is address interpolation. This method makes use of data from a street geographic information system where the street network is already mapped within the geographic coordinate space. Each street segment is attributed with address ranges (e.g. house numbers from one segment to the next). Geocoding takes an address, matches it to a street and specific segment (such as a block, in towns that use the "block" convention). Geocoding then interpolates the position of the address, within the range along the segment.

Example

Take for example: *742 Evergreen Terrace*

Let's say that this segment (for instance, a block) of Evergreen Terrace runs from 700 to 799. Even-numbered addresses fall on the east side of Evergreen Terrace, with odd-numbered addresses on the west side of the street. 742 Evergreen Terrace would (probably) be located slightly less than halfway up the block, on the east side of the street. A point would be mapped at that location along the street, perhaps offset a distance to the east of the street centerline.

Complicating Factors

However, this process is not always as straightforward as in this example. Difficulties arise when

- distinguishing between ambiguous addresses such as 742 Evergreen Terrace and 742 W Evergreen Terrace.
- attempting to geocode new addresses for a street that is not yet added to the geographic information system database.

While there might be 742 Evergreen Terrace in Springfield, there might also be a 742 Evergreen Terrace in Shelbyville. Asking for the city name (and state, province, country, etc. as needed) can solve this problem. Boston, Massachusetts has multiple "100 Washington Street" locations because several cities have been annexed without changing street names, thus requiring use of unique postal codes or district names for disambiguation. Geocoding accuracy can be greatly improved by first utilizing good address verification practices. Address verification will confirm the existence of the address and will eliminate ambiguities. Once the valid address is determined, it is very easy to geocode and determine the latitude/longitude coordinates. Finally, several caveats on using interpolation:

- The typical attribution of a street segment assumes that all even numbered parcels are on one side of the segment, and all odd numbered parcels are on the other. This is often not true in real life.
- Interpolation assumes that the given parcels are evenly distributed along the length of the segment. This is almost never true in real life; it is not uncommon for a geocoded address to be off by several thousand feet.
- Interpolation also assumes that the street is straight. If a street is curved then the geocoded location will not necessarily fit the physical location of the address.
- Segment Information (esp. from sources such as TIGER) includes a maximum upper bound for addresses and is interpolated as though the full address range is used. For example, a segment (block) might have a listed range of 100-199, but the last address at the end of the block is 110. In this case, address 110 would be geocoded to 10% of the distance down the segment rather than near the end.
- Most interpolation implementations will produce a point as their resulting address location. In reality, the physical address is distributed along the length of the segment, i.e. consider geocoding the address of a shopping mall - the physical lot may run a distance along the street segment (or could be thought of as a two-dimensional space-filling polygon which may front on several different streets — or worse, for cities with multi-level streets, a three-dimensional shape that meets different streets at several different levels) but the interpolation treats it as a singularity.

A very common error is to believe the accuracy ratings of a given map's geocodable attributes. Such accuracy currently touted by most vendors has no bearing on an address being attributed to the correct segment, being attributed to the correct side of the segment, nor resulting in an accurate position along that correct segment. With the geocoding process used for U.S. Census TIGER datasets, 5-7.5% of the addresses may be allocated to a different census tract, while a study of Australia's TIGER-like system found that 50% of the geocoded points were mapped to the wrong property parcel. The accuracy of geocoded data can also have a bearing on the quality of research that can be done using this data. One study by a group of Iowa researchers found that the common method of geocoding using TIGER datasets as described above, can cause a loss of as much as 40% of the power of a statistical analysis. An alternative is to use orthophoto or image coded data such as the Address Point data from Ordnance Survey in the UK, but such datasets are generally expensive. Because of this, it is quite important to avoid using interpolated results except for non-critical applications, such as pizza delivery. Interpolated geocoding is usually not appropriate for making authoritative decisions, for example if life safety will be affected by that decision. Emergency services, for example, do not make an authoritative decision based on their interpolations; an ambulance or fire truck will always be dispatched regardless of what the map says.

Other Techniques

In rural areas or other places lacking high quality street network data and addressing, GPS is useful for mapping a location. For traffic accidents, geocoding to a street intersection or midpoint along a street centerline is a suitable technique. Most highways in developed countries have mile markers to aid in emergency response, maintenance, and navigation. It is also possible to use a combination of these geocoding techniques – using a particular technique for certain cases and situations and other techniques for other cases. In contrast to geocoding of structured postal address records, toponym resolution maps place names in unstructured document collections to their corresponding spatial footprints.

Research

Research has introduced a new approach to the control and knowledge aspects of geocoding, by using an agent-based paradigm. In addition to the new paradigm for geocoding, additional correction techniques and control algorithms have been developed. The approach represents the geographic elements commonly found in addresses as individual agents. This provides a commonality and duality to control and geographic representation. In addition to scientific publication, the new approach and subsequent prototype gained national media coverage in Australia. The research was conducted at Curtin University in Perth, Western Australia.

Uses

Geocoded locations are useful in many GIS analysis, cartography, decision making workflow, transaction mash-up, or injected into larger business processes. On the web, geocoding is used in services like routing and local search. Geocoding, along with GPS provides location data for geotagging media, such as photographs or RSS items.

Privacy Concerns

The proliferation and ease of access to geocoding (and reverse-geocoding) services raises privacy concerns. For example, in mapping crime incidents, law enforcement agencies aim to balance the privacy rights of victims and offenders, with the public's right to know. Law enforcement agencies have experimented with alternative geocoding techniques that allow them to mask a portion of the locational detail (e.g., address specifics that would lead to identifying a victim or offender). As well, in providing online crime mapping to the public, they also place disclaimers regarding the locational accuracy of points on the map, acknowledging these location masking techniques, and impose terms of use for the information.

Geotagging

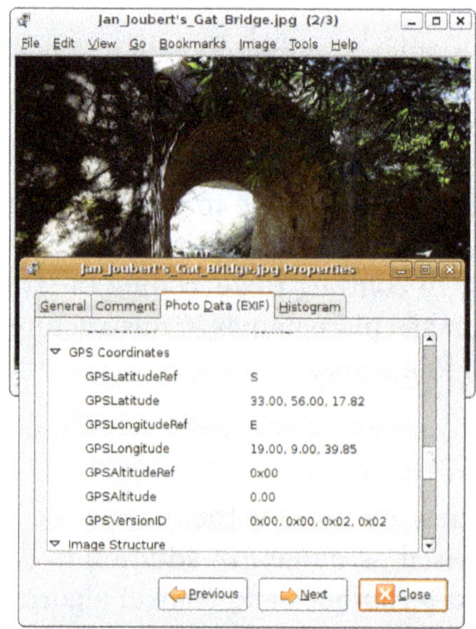

Geotag information in a JPEG photo, shown by the software gThumb

Geotag information stamped onto a JPEG photo by the software GPStamper

Geotagging (also written as GeoTagging) is the process of adding geographical identification metadata to various media such as a geotagged photograph or video, websites, SMS messages, QR Codes or RSS feeds and is a form of geospatial metadata. This data usually consists of latitude and longitude coordinates, though they can also include altitude, bearing, distance, accuracy data, and place names, and perhaps a time stamp.

Geotagging can help users find a wide variety of location-specific information from a device. For instance, someone can find images taken near a given location by entering latitude and longitude coordinates into a suitable image search engine. Geotagging-enabled information services can also potentially be used to find location-based news, websites, or other resources. Geotagging can tell users the location of the content of a given picture or other media or the point of view, and conversely on some media platforms show media relevant to a given location.

The related term geocoding refers to the process of taking non-coordinate based geographical identifiers, such as a street address, and finding associated geographic coordinates (or vice versa for reverse geocoding). Such techniques can be used together with geotagging to provide alternative search techniques.

Popular Examples

Geotagging has become a popular feature on several social media platforms, such as Facebook and Instagram.

Facebook users can geotag photos that can be added to the page of the location they are tagging. Users may also use a feature that allows them to find nearby Facebook friends, by generating a list of people according to the location tracker in their mobile devices.

Instagram uses a map feature that allows users to geotag photos. The map layout pin points specific photos that the user has taken on a world map.

Geotagging Techniques

The geographical location data used in geotagging will, in almost every case, be derived from the global positioning system, and based on a latitude/longitude-coordinate system that presents each location on the earth from 180° west through 180° east along the Equator and 90° north through 90° south along the prime meridian.

Geotagging Photos

There are two main options for geotagging photos; capturing GPS information at the time the photo is taken or "attaching" the photograph to a map after the picture is taken.

In order to capture GPS data at the time the photograph is captured, the user must have a camera with built in GPS or a standalone GPS along with a digital camera. Because of the requirement for wireless service providers in United States to supply more precise location information for 911 calls by September 11, 2012, more and more cell phones have built-in GPS chips. Most smart phones already use a GPS chip along with built-in cameras to allow users to automatically geotag photos. Others may have the GPS chip and camera but do not have internal software needed to embed the GPS information within the picture. A few digital cameras also have built-on or built-in GPS that allow for automatic geotagging. Devices use GPS, A-GPS or both. A-GPS can be faster getting an initial fix if within range of a cell phone tower, and may work better inside buildings. Traditional GPS does not need cell phone towers and uses standard GPS signals outside of urban areas. Traditional GPS tends to use more battery power. Almost any digital camera can be coupled with a stand-alone GPS and post processed with photo mapping software, to write the location information to the image's exif header.

GPS Formats

GPS coordinates may be represented in text in a number of ways, with more or fewer decimals:

Template	Description	Example
[-]d.d, [-]d.d	Decimal degrees with negative numbers for South and West.	12.3456, -98.7654
d° m.m′ {N|S}, d° m.m′ {E|W}	Degrees and decimal minutes with N, S, E or W suffix for North, South, East, West	12° 20.736′ N, 98° 45.924′ W
{N|S} d° m.m′ {E|W} d° m.m′	Degrees and decimal minutes with N, S, E or W prefix for North, South, East, West	N 12° 20.736′, W 98° 45.924′
d° m' s" {N|S}, d° m' s" {E|W}	Degrees, minutes and seconds with N, S, E or W suffix for North, South, East, West	12° 20' 44" N, 98° 45' 55" W
{N|S} d° m' s", {E|W} d° m' s"	Degrees, minutes and seconds with N, S, E or W prefix for North, South, East, West	N 12° 20' 44", W 98° 45' 55"

Geotagging Standards in Electronic file Formats

Photographs

With photos stored in JPEG, TIFF and many other file formats, the geotag information, storing camera location and sometimes heading, is typically embedded in the metadata, stored in Exchangeable image file format (Exif) or Extensible Metadata Platform (XMP) format. These data are not visible in the picture itself but are read and written by special programs and most digital cameras and modern scanners. Latitude and longitude are stored in units of degrees with decimals. This geotag information can be read by many programs, such as the cross-platform open source ExifTool. An example readout for a photo might look like:

```
GPS Latitude                    : 57 deg 38′ 56.83″ N
GPS Longitude                   : 10 deg 24′ 26.79″ E
GPS Position                  : 57 deg 38′ 56.83″ N, 10 deg 24′
26.79″ E
```

or the same coordinates could also be presented as decimal degrees:

```
GPS
GPS Longitude                          : 10.40744
GPS Position                           : 57.64911 10.40744
```

When stored in Exif, the coordinates are represented as a series of rational numbers in the GPS sub-IFD. Here is a hexadecimal dump of the relevant section of the Exif metadata (with big-endian byte order):

```
+ [GPS directory with 5 entries]
| 0)  GPSVersionID = 2 2 0 0
|     - Tag 0x0000 (4 bytes, int8u):
|         dump: 02 02 00 00
| 1)  GPSLatitudeRef = N
|     - Tag 0x0001 (2 bytes, string):
|         dump: 4e 00  [ASCII "N\0"]
| 2)  GPSLatitude = 57 38 56.83 (57/1 38/1 5683/100)
|     - Tag 0x0002 (24 bytes, rational64u):
|         dump: 00 00 00 39 00 00 00 01 00 00 00 26 00 00 00 01
|         dump: 00 00 16 33 00 00 00 64
| 3)  GPSLongitudeRef = W
|     - Tag 0x0003 (2 bytes, string):
|         dump: 57 00  [ASCII "W\0"]
| 4)  GPSLongitude = 10 24 26.79 (10/1 24/1 2679/100)
|     - Tag 0x0004 (24 bytes, rational64u):
|         dump: 00 00 00 0a 00 00 00 01 00 00 00 18 00 00 00 01
|         dump: 00 00 0a 77 00 00 00 64
```

Remote Sensing Data

In the field of remote sensing the geotagging goal is to store coordinates of every pixel in the image. One approach is used with the orthophotos where we store coordinates of four corners and all the other pixels can be georeferenced by interpolation. The four corners are stored using GeoTIFF or World file standards. Hyperspectral images take a different approach defining a separate file of the same spatial dimensions as the image where latitude and longitude of each pixel are stored as two 2D layers in so called *Input geometry data* (IGM) files, also known as GEO files.

Audio/Video Files

Audio/video files can be geotagged via: metadata, audio encoding, overlay, or with companion files. Metadata records the geospatial data in the encoded video file to be decoded for later analysis. One of the standards used with unmanned aerial vehicle is MISB Standard 0601 which allows geocoding of corner points and horizon lines in individual video frames. Audio encoding involves a process of converting gps data into audio data such as modem squawk. Overlay involves overlaying GPS data as text on the recorded video. Companion files are separate data files which correspond to respective audio/video files. Companion files are typically found in the .KML and .GPX data formats. For audio and video files which use the vorbiscomment metadata format (including Opus, Ogg Vorbis, FLAC, Speex, and Ogg Theora), there is a proposed GEO_LOCATION field which can be used. Like all vorbiscomments, it is plain text, and it takes the form:

```
GEO_LOCATION=(decimal latitude);(decimal longitude);([optional] elevation in meters)
```

for example:

```
GEO_LOCATION=35.1592;-98.4422;410
```

SMS Messages

The GeoSMS standard works by embedding one or more 'geo' URIs in the body of an SMS, for example:

```
I'm at the pub geo:-37.801631,144.980294;u=16
```

DNS Entries

RFC 1876 defines a means for expressing location information in the Domain Name System. LOC resources records can specify the latitude, longitude, altitude, precision of the location, and the physical size of on entity attached to an IP address. However, in practice not all IP addresses have such a record, so it is more common to use geolocation services to find the physical location of an IP address.

HTML Pages

ICBM Method

The GeoURL method requires the ICBM tag (plus optional Dublin Core metadata), which is used to geotag standard web pages in HTML format:

```
<meta name="ICBM" content="50.167958, -97.133185">
```

The similar Geo Tag format allows the addition of placename and region tags:

```
<meta name="geo.position" content="50.167958;-97.133185">

<meta name="geo.placename" content="Rockwood Rural Municipali-
ty, Manitoba, Canada">

<meta name="geo.region" content="ca-mb">
```

RDF Feeds

The RDF method is defined by W3 Group and presents the information in RDF tags:

```
<rdf:RDF     xmlns:rdf="http://www.w3.org/1999/02/22-rdf-syntax-
ns#"

         xmlns:geo="http://www.w3.org/2003/01/geo/wgs84_pos#">

  <geo:Point>

  <geo:lat>55.701</geo:lat>

  <geo:long>12.552</geo:long>

  </geo:Point>

</rdf:RDF>
```

Microformat

The Geo microformat allows coordinates within HyperText Markup Language pages to be marked up in such a way that they can be "discovered" by software tools. Example:

```
<span class="geo">

   <span class="latitude">50.167958</span>;

   <span class="longitude">-97.133185</span>

</span>
```

A proposal has been developed to extend Geo to cover other bodies, such as Mars and the Moon.

An example is the Flickr photo-sharing Web site, which provides geographic data for any geotagged photo in all of the above-mentioned formats.

Geotagging in tag-based Systems

No industry standards exist, however there are a variety of techniques for adding geographical identification metadata to an information resource. One convention, established by the website Geobloggers and used by more and more sites, e.g. photo sharing sites Panoramio and Flickr, and the social bookmarking site del.icio.us, enables content to be found via a location search. Such sites allow users to add metadata to an information resource via a set of so-called *machine tags*.

```
geotagged

geo:lat=57.64911

geo:lon=10.40744
```

This describes the geographic coordinates of a particular location in terms of latitude (geo:lat) and longitude (geo:lon). These are expressed in decimal degrees in the WGS84 datum, which has become something of a default geodetic datum with the advent of GPS.

Using three tags works within the constraint of having tags that can only be single 'words'. Identifying geotagged information resources on sites like Flickr and del.icio.us is done by searching for the 'geotagged' tag, since the tags beginning 'geo:lat=' and 'geo:lon=' are necessarily very variable.

Another option is to tag with a Geohash:

```
geo:hash=u4pruydqqvj
```

A further convention proposed by FlickrFly adds tags to specify the suggested viewing angle and range when the geotagged location is viewed in Google Earth:

```
ge:head=225.00

ge:tilt=45.00

ge:range=560.00
```

These three tags would indicate that the camera is pointed heading 225° (south west), has a 45° tilt and is 560 metres from the subject.

Where the above methods are in use, their coordinates may differ from those specified by the photo's internal Exif data, for example because of a correction or a difference between the camera's location and the subject's.

In order to integrate geotags in social media and enhance text readability or oral use, the concept of 'meetag' or tag-to-meet has been proposed. Differing from hashtag construction, meetag includes the geolocation information after an underscore. A meetag is therefore a word or an unspaced phrase prefixed with an underscore ("_"). Words in messages on microblogging and social networking services may be tagged by putting "_" before them, either as they appear in a sentence, (e.g. "There is a concert going _montreuxjazzfestival", "the world wide web was invented _cern _geneve", ...) or appended to it.

Geoblogging

Geoblogging attaches specific geographic location information to blog entries via *geotags*. Searching a list of blogs and pictures tagged using geotag technology allows users to select areas of specific interest to them on interactive maps.

The progression of GPS technology, along with the development of various online applications, has fueled the popularity of such tagged blogging, and the combination of GPS Phones and GSM localization, has led to the moblogging, where blog posts are tagged with exact position of the user. Real-time geotagging relays automatically geotagged media such as photos or video to be published and shared immediately.

For better integration and readability of geotags into blog texts, the meetag syntax has been proposed, which transforms any word, sentence, or precise geolocalization coordinates prefixed with an underscore into a 'meetag'. It not only lets one express a precise location but also takes in account dynamically changing geolocations.

Wikipedia Article Geosearching Apps

One of the first attempts to initiate the geotagging aspect of searching and locating articles seems to be the now-inoperative site Wikinear.com, launched in 2008, which showed the user Wikipedia pages that are geographically closest to one's current location.

The 2009 app Cyclopedia works relatively well showing geotagged Wikipedia articles located within several miles of ones location, integrated with a street-view mode, and 360-degree mode.

The app Respotter Wiki, launched in 2009, claims to feature Wikipedia searching via a map, also allowing users to interact with people around them, via messaging and reviews, etc. The app, in its current function, however, seems to give only geotagged photo results.

Dangers of Geotagging

Following a scientific study and several demonstrative websites, a discussion on the privacy implications of geotagging has raised public attention. In particular, the automatic embedding of geotags in pictures taken with smartphones is often ignored by cell-phone users. As a result, people are often not aware that the photos they publish on the Internet have been geotagged. Many celebrities reportedly gave away their home location without knowing it. According to the study, a significant number of for-sale advertisements on Craigslist, that were otherwise anonymized, contained geotags, thereby revealing the location of high-valued goods—sometimes in combination with clear hints to the absence of the offerer at certain times. Publishing photos and other media tagged with exact geolocation on the Internet allows random people to track an individual's location and correlate it with other information. Therefore, criminals could find out when homes are empty because their inhabitants posted geotagged and timestamped information both about their home address and their vacation residence. These dangers can be avoided by removing geotags with a metadata removal tool for photos before publishing them on the Internet.

In 2007, four United States Army Apache helicopters were destroyed on the ground by Iraqi insurgent mortar fire; the insurgents had made use of embedded coordinates in web-published photographs (geotagging) taken of the helicopters by soldiers.

Another newly realised danger of geotagging is the location information provided to criminal gangs and poachers on the whereabouts of often endangered animals. This can effectively make tourists scouts for these poachers, so geotagging should be turned off when photographing these animals.

Reverse Geocoding

Reverse geocoding is the process of back (reverse) coding of a point location (latitude, longitude) to a readable address or place name. This permits the identification of nearby street addresses, places, and/or areal subdivisions such as neighbourhoods, county, state, or country. Combined with geocoding and routing services, reverse geocoding is a critical component of mobile location-based services and Enhanced 911 to convert a coordinate obtained by GPS to a readable street address which is easier to understand by the end user.

Reverse geocoding can be carried out systematically by services which process a coordinate similarly to the geocoding process. For example, when a GPS coordinate is entered the street address is interpolated from a range assigned to the road segment in a reference dataset that the point is nearest to. If the user provides a coordinate near the midpoint of a segment that starts with address 1 and ends with 100, the returned street address will be somewhere near 50. This approach to reverse geocoding does not return actual addresses, only estimates of what should be there based on the predetermined range. Alternatively, coordinates for reverse geocoding can also be selected on an interactive map, or extracted from static maps by georeferencing them in a GIS with predefined spatial layers to determine the coordinates of a displayed point. Many of the same limitations of geocoding are similar with reverse geocoding.

Public reverse geocoding services are becoming increasingly available through APIs and other web services as well as mobile phone applications. These services require manual input of a coordinate, capture from a localization tool (mostly GPS, but also cell tower signals or WiFi traces), or selection of a point on an interactive map; to look up a street address or neighboring places. Examples of these services include the GeoNames reverse geocoding web service which has tools to identify nearest street address, place names, Wikipedia articles, country, county subdivisions, neighborhoods, and other location data from a coordinate. Google has also published a reverse geocoding API which can be adapted for online reverse geocoding tools, which uses the same street reference layer as Google maps.

Privacy Concerns

Geocoding and reverse geocoding have raised potential privacy concerns, especially regarding the ability to reverse engineer street addresses from published static maps. By digitizing published maps it is possible to georeference them by overlaying with other spatial layers and then extract point locations which can be used to identify individuals or reverse geocoded to obtain a street address of the individual. This has potential implications to determine locations for patients or study participants from maps published in medical literature as well as potentially sensitive information published in other journalistic sources.

In one study a map of Hurricane Katrina mortality locations published in a Baton Rouge, Louisiana, paper was examined. Using GPS locations obtained from houses where fatalities occurred, the authors were able to determine the relative error between the true house locations and the location determined by georeferencing the published map. The authors found that approximately 45% of the points extracted from the georeferenced map were within 10 meters of a household's GPS obtained point. Another study found similar results in examining hypothetical low and high-resolution patient address maps similar to what might be found published in medical journals. They found approximately 26% of points obtained from a low-resolution map and 79% from a high-resolution map were matched precisely with the true location.

The findings from these studies raise concerns regarding the potential use of georeferencing and reverse geocoding of published maps to elucidate sensitive or private information on mapped individuals. Guidelines for the display and publication of potentially sensitive information are inconsistently applied and no uniform procedure has been identified. The use of blurring algorithms which shift the location of mapped points have been proposed as a solution. In addition, where direct reference to the geography of the area mapped is not required, it may be possible to use abstract space on which to display spatial patterns.

Geographic Coordinate System

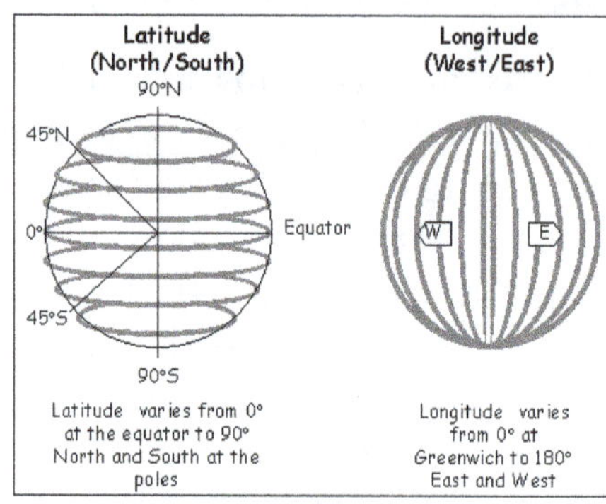

Longitude lines are perpendicular and latitude lines are parallel to the equator.

A geographic coordinate system is a coordinate system that enables every location on the Earth to be specified by a set of numbers, letters or symbols.[n 1] The coordinates are often chosen such that one of the numbers represents vertical position, and two or three of the numbers represent horizontal position. A common choice of coordinates is latitude, longitude and elevation.

To specify a location on a two-dimensional map requires a map projection.

History

The invention of a geographic coordinate system is generally credited to Eratosthenes of Cyrene, who composed his now-lost *Geography* at the Library of Alexandria in the 3rd century BC. A century later, Hipparchus of Nicaea improved upon his system by determining latitude from stellar measurements rather than solar altitude and determining longitude by using simultaneous timing of lunar eclipses, rather than dead reckoning. In the 1st or 2nd century, Marinus of Tyre compiled an extensive gazetteer and mathematically-plotted world map, using coordinates measured east from a Prime Meridian at the Fortunate Isles of western Africa and measured north or south of the island of Rhodes off Asia Minor. Ptolemy credited him with the full adoption of longitude and latitude, rather than measuring latitude in terms of the length of the midsummer day. Ptolemy's 2nd-century *Geography* used the same Prime Meridian but measured latitude from the equator instead. After their work was translated into Arabic in the 9th century, Al-Khwārizmī's *Book of the Description of the Earth* corrected Marinus and Ptolemy's errors regarding the length of the Mediterranean Sea,[n 2] causing medieval Arabic cartography to use a Prime Meridian around 10° east of Ptolemy's line. Mathematical cartography resumed in Europe following Maximus Planudes's recovery of Ptolemy's text a little before 1300; the text was translated into Latin at Florence by Jacobus Angelus around 1407.

In 1884, the United States hosted the International Meridian Conference and twenty-five nations attended. Twenty-two of them agreed to adopt the longitude of the Royal Observatory in Greenwich, England, as the zero-reference line. The Dominican Republic voted against the motion, while France and Brazil abstained. France adopted Greenwich Mean Time in place of local determinations by the Paris Observatory in 1911.

Geographic Latitude and Longitude

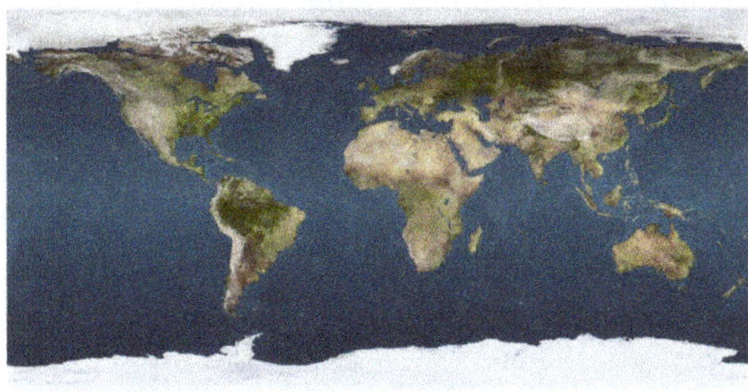

0°

Equator

The "latitude" (abbreviation: Lat., φ, or phi) of a point on the Earth's surface is the angle between the equatorial plane and the straight line that passes through that point and through (or close to) the center of the Earth.[n 3] Lines joining points of the same latitude trace circles on the surface of the Earth called parallels, as they are parallel to the equator and to each other. The north pole is 90° N; the south pole is 90° S. The 0° parallel of latitude is designated the equator, the fundamental plane of all geographic coordinate systems. The equator divides the globe into Northern and Southern Hemispheres.

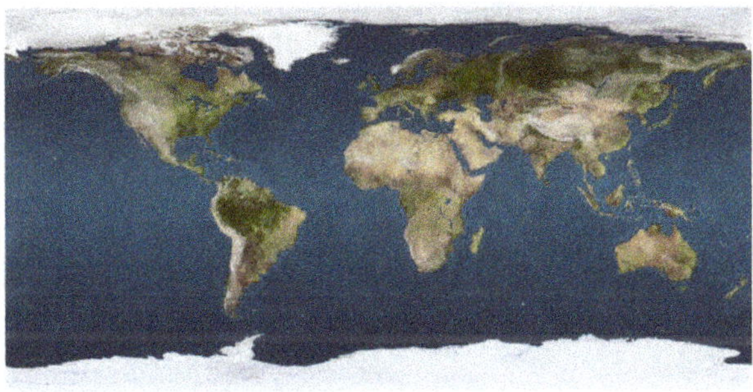

0°

Prime Meridian

The "longitude" (abbreviation: Long., λ, or lambda) of a point on the Earth's surface is the angle east or west from a reference meridian to another meridian that passes through that point. All meridians are halves of great ellipses (often improperly called great circles), which converge at the north and south poles. The meridian of the British Royal Observatory in Greenwich, in south-east London, England, is the international Prime Meridian although some organizations—such as the French Institut Géographique National—continue to use other meridians for internal purposes. The Prime Meridian determines the proper Eastern and Western Hemispheres, although maps often divide these hemispheres further west in order to keep the Old World on a single side. The antipodal meridian of Greenwich is both 180°W and 180°E. This is not to be conflated with the International Date Line, which diverges from it in several places for political reasons including between far eastern Russia and the far western Aleutian Islands.

The combination of these two components specifies the position of any location on the surface of the Earth, without consideration of altitude or depth. The grid thus formed by latitude and longitude is known as the "graticule". The zero/zero point of this system is located in the Gulf of Guinea about 625 km (390 mi) south of Tema, Ghana.

Measuring Height using Datums

Complexity of The Problem

To completely specify a location of a topographical feature on, in, or above the Earth, one has to also specify the vertical distance from the center of the Earth, or from the surface of the Earth.

The Earth is not a sphere, but an irregular shape approximating a biaxial ellipsoid. It is nearly spherical, but has an equatorial bulge making the radius at the equator about 0.3% larger than the radius measured through the poles. The shorter axis approximately coincides with axis of rotation. Though early navigators thought of the sea as a flat surface that could be used as a vertical datum, this is not actually the case. The Earth has a series of layers of equal potential energy within its gravitational field. Height is a measurement at right angles to this surface, roughly toward the centre of the Earth, but local variations make the equipotential layers irregular (though roughly ellipsoidal). The choice of which layer to use for defining height is arbitrary.

Common Baselines

Common height baselines include

- The surface of the datum ellipsoid, resulting in an *ellipsoidal height*
- The mean sea level as described by the gravity geoid, yielding the orthometric height
- A vertical datum, yielding a dynamic height relative to a known reference height.

Along with the latitude ϕ and longitude λ, the height h provides the three-dimensional *geodetic coordinates* or *geographic coordinates* for a location.

Datums

In order to be unambiguous about the direction of "vertical" and the "surface" above which they are measuring, map-makers choose a reference ellipsoid with a given origin and orientation that best fits their need for the area they are mapping. They then choose the most appropriate mapping of the spherical coordinate system onto that ellipsoid, called a *terrestrial reference system* or geodetic datum.

Datums may be *global*, meaning that they represent the whole earth, or they may be *local*, meaning that they represent a best-fit ellipsoid to only a portion of the earth. Points on the Earth's surface move relative to each other due to continental plate motion, subsidence, and diurnal movement caused by the Moon and the tides. The daily movement can be as much as a metre. Continental movement can be up to 10 cm a year, or 10 m in a century. A weather system high-pressure area can cause a sinking of 5 mm. Scandinavia is rising by 1 cm a year as a result of the melting of the ice sheets of the last ice age, but neighbouring Scotland is rising by only 0.2 cm. These changes are insignificant if a local datum is used, but are statistically significant if a global datum is used.

Examples of global datums include World Geodetic System (WGS 84), the default datum used for Global Positioning System and the International Terrestrial Reference Frame (ITRF) used for estimating continental drift and crustal deformation. The distance to Earth's centre can be used both for very deep positions and for positions in space.

Local datums chosen by a national cartographical organisation include the North American Datum, the European ED50, and the British OSGB36. Given a location, the datum provides the latitude ϕ and longitude λ. In the United Kingdom there are three common latitude, longitude, height systems in use. WGS 84 differs at Greenwich from the one used on published maps OSGB36 by approximately 112m. The military system ED50, used by NATO, differs by about 120m to 180m.

The latitude and longitude on a map made against a local datum may not be the same as on a GPS receiver. Coordinates from the mapping system can sometimes be roughly changed into another datum using a simple translation. For example, to convert from ETRF89 (GPS) to the Irish Grid add 49 metres to the east, and subtract 23.4 metres from the north. More generally one datum is changed into any other datum using a process called Helmert transformations. This involves converting the spherical coordinates into Cartesian coordinates and applying a seven parameter transformation (translation, three-dimensional rotation), and converting back.

In popular GIS software, data projected in latitude/longitude is often represented as a 'Geographic Coordinate System'. For example, data in latitude/longitude if the datum is the North American Datum of 1983 is denoted by 'GCS North American 1983'.

Map Projection

To establish the position of a geographic location on a map, a map projection is used to convert geodetic coordinates to two-dimensional coordinates on a map; it projects the datum ellipsoidal coordinates and height onto a flat surface of a map. The datum, along with a map projection applied to a grid of reference locations, establishes a *grid system* for plotting locations. Common map projections in current use include the Universal Transverse Mercator (UTM), the Military grid reference system (MGRS), the United States National Grid (USNG), the Global Area Reference System (GARS) and the World Geographic Reference System (GEOREF). Coordinates on a map are usually in terms northing N and easting E offsets relative to a specified origin.

Map projection formulas depend in the geometry of the projection as well as parameters dependent on the particular location at which the map is projected. The set of parameters can vary based on type of project and the conventions chosen for the projection. For the transverse Mercator projection used in UTM, the parameters associated are the latitude and longitude of the natural origin, the false northing and false easting, and an overall scale factor. Given the parameters associated with particular location or grin, the projection formulas for the transverse Mercator are a complex mix of algebraic and trigonometric functions.

UTM and UPS Systems

The Universal Transverse Mercator (UTM) and Universal Polar Stereographic (UPS) coordinate systems both use a metric-based cartesian grid laid out on a conformally projected surface to locate positions on the surface of the Earth. The UTM system is not a single map projection but a series of sixty, each covering 6-degree bands of longitude. The UPS system is used for the polar regions, which are not covered by the UTM system.

Stereographic Coordinate System

During medieval times, the stereographic coordinate system was used for navigation purposes. The stereographic coordinate system was superseded by the latitude-longitude system. Although no longer used in navigation, the stereographic coordinate system is still used in modern times to describe crystallographic orientations in the fields of crystallography, mineralogy and materials science.

Cartesian Coordinates

Every point that is expressed in ellipsoidal coordinates can be expressed as an rectilinear x y z (Cartesian) coordinate. Cartesian coordinates simplify many mathematical calculations. The Cartesian systems of different datums are not equivalent.

Earth-Centered, Earth-Fixed

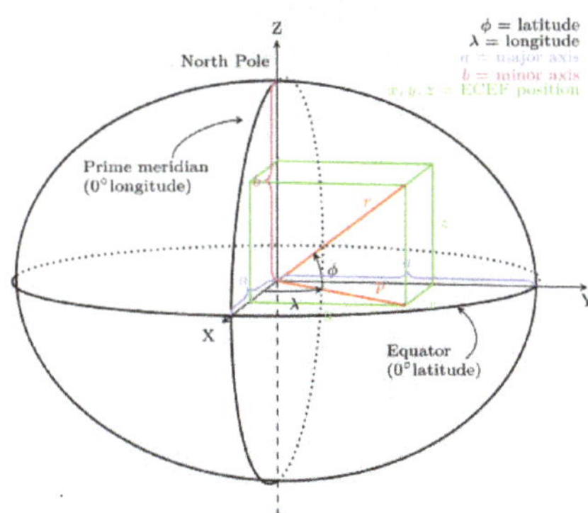

Earth Centered, Earth Fixed coordinates in relation to latitude and longitude.

The earth-centered earth-fixed (also known as the ECEF, ECF, or conventional terrestrial coordinate system) rotates with the Earth and has its origin at the center of the Earth.

The conventional right-handed coordinate system puts:

- The origin at the center of mass of the earth, a point close to the Earth's center of figure
- The Z axis on the line between the north and south poles, with positive values increasing northward (but does not exactly coincide with the Earth's rotational axis)
- The X and Y axes in the plane of the equator
- The X axis passing through extending from 180 degrees longitude at the equator (negative) to 0 degrees longitude (prime meridian) at the equator (positive)
- The Y axis passing through extending from 90 degrees west longitude at the equator (negative) to 90 degrees east longitude at the equator (positive)

An example is the NGS data for a brass disk near Donner Summit, in California. Given the dimensions of the ellipsoid, the conversion from lat/lon/height-above-ellipsoid coordinates to X-Y-Z is straightforward—calculate the X-Y-Z for the given lat-lon on the surface of the ellipsoid and add the X-Y-Z vector that is perpendicular to the ellipsoid there and has length equal to the point's height above the ellipsoid. The reverse conversion is harder: given X-Y-Z we can immediately get longitude, but no closed formula for latitude and height exists." Using Bowring's formula in 1976 *Survey Review* the first iteration gives latitude correct within 10^{-11} degree as long as the point is within 10000 meters above or 5000 meters below the ellipsoid.

Local East, North, Up (ENU) Coordinates

In many targeting and tracking applications the local East, North, Up (ENU) Cartesian coordinate system is far more intuitive and practical than ECEF or Geodetic coordinates. The local ENU coordinates are formed from a plane tangent to the Earth's surface fixed to a specific location and hence it is sometimes known as a "Local Tangent" or "local geodetic" plane. By convention the east axis is labeled x, the north y and the up z.

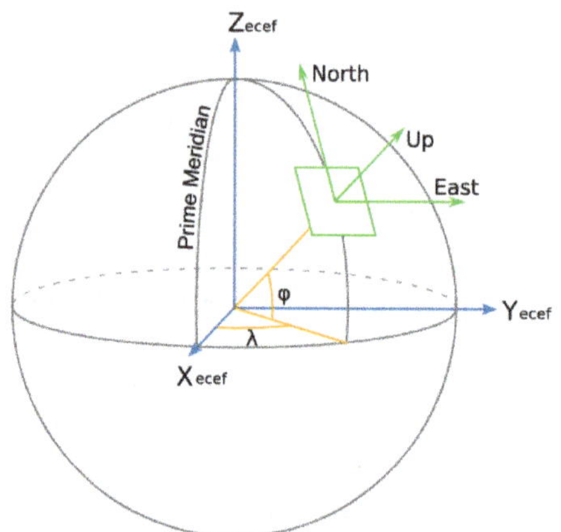

Earth Centred Earth Fixed and East, North, Up coordinates.

Local North, East, Down (NED) Coordinates

Also known as local tangent plane (LTP). In an airplane, most objects of interest are below the aircraft, so it is sensible to define down as a positive number. The North, East, Down (NED) coordinates allow this as an alternative to the ENU local tangent plane. By convention, the north axis is labeled x', the east y' and the down z'. To avoid confusion between x and x', etc. in this web page we will restrict the local coordinate frame to ENU.

Expressing Latitude and Longitude as Linear Units

On the GRS80 or WGS84 spheroid at sea level at the equator, one latitudinal second measures *30.715 metres*, one latitudinal minute is *1843 metres* and one latitudinal degree is *110.6 kilometres*. The circles of longitude, meridians, meet at the geographical poles, with the west-east width of a second naturally decreasing as latitude increases. On the equator at sea level, one longitudinal second measures *30.92 metres*, a longitudinal minute is *1855 metres* and a longitudinal degree is *111.3 kilometres*. At 30° a longitudinal second is *26.76 metres*, at Greenwich (51°28′38″N) *19.22 metres*, and at 60° it is *15.42 metres*.

On the WGS84 spheroid, the length in meters of a degree of latitude at latitude φ (that is, the distance along a north-south line from latitude (φ – 0.5) degrees to (φ + 0.5) degrees) is about

$$111132.92 - 559.82\cos 2\varphi + 1.175\cos 4\varphi - 0.0023\cos 6\varphi$$

Similarly, the length in meters of a degree of longitude can be calculated as

$$111412.84\cos\varphi - 93.5\cos 3\varphi + 0.118\cos 5\varphi$$

(Those coefficients can be improved, but as they stand the distance they give is correct within a centimeter.)

An alternative method to estimate the length of a longitudinal degree at latitude φ is to assume a spherical Earth (to get the width per minute and second, divide by 60 and 3600, respectively):

$$\frac{\pi}{180} M_r \cos\varphi$$

where Earth's average meridional radius M_r is 6,367,449 m. Since the Earth is not spherical that result can be off by several tenths of a percent; a better approximation of a longitudinal degree at latitude φ is

$$\frac{\pi}{180} a \cos\beta$$

where Earth's equatorial radius a equals *6,378,137 m* and $\tan\beta = \frac{b}{a}\tan\varphi$;; for the GRS80 and WGS84 spheroids, b/a calculates to be 0.99664719. (β is known as the reduced (or parametric) latitude). Aside from rounding, this is the exact distance along a parallel of latitude; getting the distance along the shortest route will be more work, but those two distances are always within 0.6 meter of each other if the two points are one degree of longitude apart.

Longitudinal length equivalents at selected latitudes					
Latitude	**City**	**Degree**	**Minute**	**Second**	**±0.0001°**
60°	Saint Petersburg	55.80 km	0.930 km	15.50 m	5.58 m
51° 28′ 38″ N	Greenwich	69.47 km	1.158 km	19.30 m	6.95 m
45°	Bordeaux	78.85 km	1.31 km	21.90 m	7.89 m
30°	New Orleans	96.49 km	1.61 km	26.80 m	9.65 m
0°	Quito	111.3 km	1.855 km	30.92 m	11.13 m

Geostationary Coordinates

Geostationary satellites (e.g., television satellites) are over the equator at a specific point on Earth, so their position related to Earth is expressed in longitude degrees only. Their latitude is always zero (or approximately so), that is, over the equator.

On other Celestial Bodies

Similar coordinate systems are defined for other celestial bodies such as:

- A similarly well-defined system based on the reference ellipsoid for Mars.
- Selenographic coordinates for the Moon

Types of Geographic Coordinates

Longitude

Longitude , is a geographic coordinate that specifies the east-west position of a point on the Earth's surface. It is an angular measurement, usually expressed in degrees and denoted by the Greek letter lambda (λ). Meridians (lines running from the North Pole to the South Pole) connect points with the same longitude. By convention, one of these, the Prime Meridian, which passes through the Royal Observatory, Greenwich, England, was allocated the position of zero degrees longitude. The longitude of other places is measured as the angle east or west from the Prime Meridian, ranging from 0° at the Prime Meridian to +180° eastward and −180° westward. Specifically, it is the angle between a plane containing the Prime Meridian and a plane containing the North Pole, South Pole and the location in question. (This forms a right-handed coordinate system with the *z* axis (right hand thumb) pointing from the Earth's center toward the North Pole and the *x* axis (right hand index finger) extending from Earth's center through the equator at the Prime Meridian.)

A location's north–south position along a meridian is given by its latitude, which is approximately the angle between the local vertical and the plane of the Equator.

If the Earth were perfectly spherical and homogeneous, then the longitude at a point would be equal to the angle between a vertical north–south plane through that point and the plane of the Greenwich meridian. Everywhere on Earth the vertical north–south plane would contain the Earth's axis. But the Earth is not homogeneous, and has mountains—which have gravity and so can shift the vertical plane away from the Earth's axis. The vertical north–south plane still intersects the plane of the Greenwich meridian at some angle; that angle is the astronomical longitude, calculated from star observations. The longitude shown on maps and GPS devices is the angle between the Greenwich plane and a not-quite-vertical plane through the point; the not-quite-vertical plane is perpendicular to the surface of the spheroid chosen to approximate the Earth's sea-level surface, rather than perpendicular to the sea-level surface itself.

History

The measurement of longitude is important both to cartography and for ocean navigation. Mariners and explorers for most of history struggled to determine longitude. Finding a method of determining longitude took centuries, resulting in the history of longitude recording the effort of some of the greatest scientific minds.

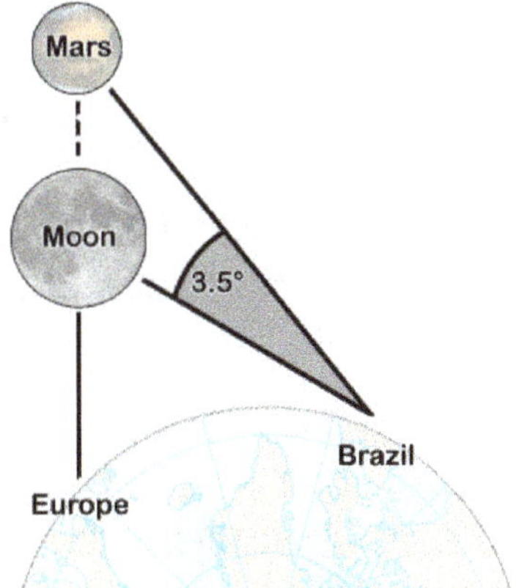

Amerigo Vespucci's means of determining longitude

Latitude was calculated by observing with quadrant or astrolabe the altitude of the sun or of charted stars above the horizon, but longitude is harder.

Amerigo Vespucci was perhaps the first European to proffer a solution, after devoting a great deal of time and energy studying the problem during his sojourns in the New World:

As to longitude, I declare that I found so much difficulty in determining it that I was put to great pains to ascertain the east-west distance I had covered. The final result of my labours was that I found nothing better to do than to watch for and take observations at night of the conjunction of one planet with another, and especially of the conjunction of the moon with the other planets, because the moon is swifter in her course than any other planet. I compared my observations with an almanac. After I had made experiments many nights, one night, the twenty-third of August 1499, there was a conjunction of the moon with Mars, which according to the almanac was to occur at midnight or a half hour before. I found that...at midnight Mars's position was three and a half degrees to the east.

John Harrison solved the greatest problem of his day.

By comparing the positions of the moon and Mars with their anticipated positions, Vespucci was able to crudely deduce his longitude. But this method had several limitations: First, it required the occurrence of a specific astronomical event (in this case, Mars passing through the same right ascension as the moon), and the observer needed to anticipate this event via an astronomical almanac. One needed also to know the precise time, which was difficult to ascertain in foreign lands. Finally, it required a stable viewing platform, rendering the technique useless on the rolling deck of a ship at sea.

In 1612 Galileo Galilei demonstrated that with sufficiently accurate knowledge of the orbits of the moons of Jupiter one could use their positions as a universal clock and this would make possible the determination of longitude, but the method he devised was impracticable for navigators on ships because of their instability. In 1714 the British government passed the Longitude Act which offered large financial rewards to the first person to demonstrate a practical method for determining the longitude of a ship at sea. These rewards motivated many to search for a solution.

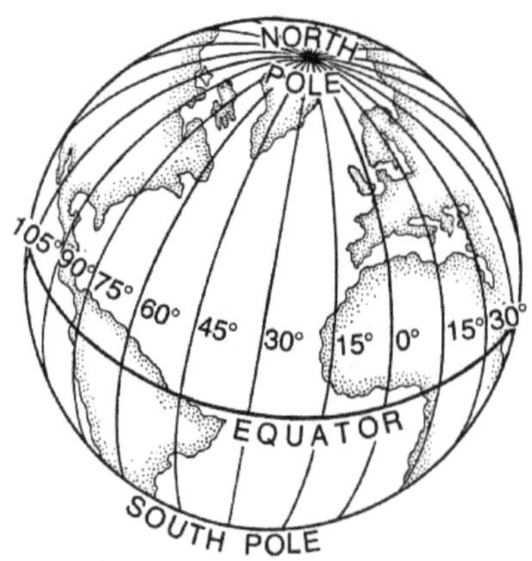

Drawing of Earth with longitudes but without latitudes.

John Harrison, a self-educated English clockmaker, invented the marine chronometer, the key piece in solving the problem of accurately establishing longitude at sea, thus revolutionising and extending the possibility of safe long distance sea travel. Though the Board of Longitude rewarded John Harrison for his marine chronometer in 1773, chronometers remained very expensive and the lunar distance method continued to be used for decades. Finally, the combination of the availability of marine chronometers and wireless telegraph time signals put an end to the use of lunars in the 20th century.

Unlike latitude, which has the equator as a natural starting position, there is no natural starting position for longitude. Therefore, a reference meridian had to be chosen. It was a popular practice to use a nation's capital as the starting point, but other locations were also used. While British cartographers had long used the Greenwich meridian in London, other references were used elsewhere, including El Hierro, Rome, Copenhagen, Jerusalem, Saint Petersburg, Pisa, Paris, Philadelphia, and Washington D.C. In 1884 the International Meridian Conference adopted the Greenwich meridian as the *universal Prime Meridian* or *zero point of longitude.*

Noting and Calculating Longitude

Longitude is given as an angular measurement ranging from 0° at the Prime Meridian to +180° eastward and −180° westward. The Greek letter λ (lambda), is used to denote the location of a place on Earth east or west of the Prime Meridian.

Each degree of longitude is sub-divided into 60 minutes, each of which is divided into 60 seconds. A longitude is thus specified in sexagesimal notation as 23° 27′ 30″ E. For higher precision, the seconds are specified with a decimal fraction. An alternative representation uses degrees and minutes, where parts of a minute are expressed in decimal notation with a fraction, thus: 23° 27.5′ E. Degrees may also be expressed as a decimal fraction: 23.45833° E. For calculations, the angular measure may be converted to radians, so longitude may also be expressed in this manner as a signed fraction of π (pi), or an unsigned fraction of 2π.

For calculations, the West/East suffix is replaced by a negative sign in the western hemisphere. Confusingly, the convention of negative for East is also sometimes seen. The preferred convention—that East is positive—is consistent with a right-handed Cartesian coordinate system, with the North Pole up. A specific longitude may then be combined with a specific latitude (usually positive in the northern hemisphere) to give a precise position on the Earth's surface.

There is no other physical principle determining longitude directly but with time. Longitude at a point may be determined by calculating the time difference between that at its location and Coordinated Universal Time (UTC). Since there are 24 hours in a day and 360 degrees in a circle, the sun moves across the sky at a rate of 15 degrees per hour (360° ÷ 24 hours = 15° per hour). So if the time zone a person is in is three hours ahead of UTC then that person is near 45° longitude (3 hours × 15° per hour = 45°). The word *near* was used because the point might not be at the center of the time zone; also the time zones are defined politically, so their centers and boundaries often do not lie on meridians at multiples of 15°. In order to perform this calculation, however, a person needs to have a chronometer (watch) set to UTC and needs to determine local time by solar or astronomical observation. The details are more complex than described here:

Singularity and Discontinuity of Longitude

Note that the longitude is singular at the Poles and calculations that are sufficiently accurate for other positions, may be inaccurate at or near the Poles. Also the discontinuity at the ±180° meridian must be handled with care in calculations. An example is a calculation of east displacement by subtracting two longitudes, which gives the wrong answer if the two positions are on either side of this meridian. To avoid these complexities, consider replacing latitude and longitude with another horizontal position representation in calculation.

Plate Movement and Longitude

The Earth's tectonic plates move relative to one another in different directions at speeds on the order of 50 to 100mm per year. So points on the Earth's surface on different plates are always in motion relative to one another, for example, the longitudinal difference between a point on the Equator in Uganda, on the African Plate, and a point on the Equator in Ecuador, on the South American Plate, is increasing by about 0.0014 arcseconds per year. These tectonic movements likewise affect latitude.

If a global reference frame (such as WGS84, for example) is used, the longitude of a place on the surface will change from year to year. To minimize this change, when dealing just with points on a single plate, a different reference frame can be used, whose coordinates are fixed to a particular plate, such as "NAD83" for North America or "ETRS89" for Europe.

Length of a Degree of Longitude

The length of a degree of longitude (east-west distance) depends only on the radius of a circle of latitude. For a sphere of radius a that radius at latitude φ is $a \cos \varphi$, and the length of a one-degree (or $\pi/180$ radian) arc along a circle of latitude is

$$\Delta^1_{long} = \frac{\pi}{180^\circ} a \cos \phi$$

φ	**Δ1 lat**	**Δ1 long**
0°	110.574 km	111.320 km
15°	110.649 km	107.551 km
30°	110.852 km	96.486 km
45°	111.132 km	78.847 km
60°	111.412 km	55.800 km
75°	111.618 km	28.902 km
90°	111.694 km	0.000 km

When the Earth is modelled by an ellipsoid this arc length becomes

$$\Delta^1_{long} = \frac{\pi a \cos \phi}{180^\circ \sqrt{1 - e^2 \sin^2 \phi}}$$

where e, the eccentricity of the ellipsoid, is related to the major and minor axes (the equatorial and polar radii respectively) by

$$e^2 = \frac{a^2 - b^2}{a^2}$$

An alternative formula is

$$\Delta^1_{long} = \frac{\pi}{180^\circ} \text{a} \cos\psi \quad \textit{where} \ \tan\psi = \frac{\text{b}}{\text{a}} \tan\phi$$

Cos φ decreases from 1 at the equator to 0 at the poles, which measures how circles of latitude shrink from the equator to a point at the pole, so the length of a degree of longitude decreases likewise. This contrasts with the small (1%) increase in the length of a degree of latitude (north-south distance), equator to pole. The table shows both for the WGS84 ellipsoid with a = 6378137.0 m and b = 6356752.3142 m. Note that the distance between two points 1 degree apart on the same circle of latitude, measured along that circle of latitude, is slightly more than the shortest (geodesic) distance between those points (unless on the equator, where these are equal); the difference is less than 0.6 m (2 ft).

A geographical mile is defined to be the length of one minute of arc along the equator (one equatorial minute of longitude), so a degree of longitude along the equator is exactly 60 geographical miles, as there are 60 minutes in a degree.

Longitude on Bodies other than Earth

Planetary co-ordinate systems are defined relative to their mean axis of rotation and various definitions of longitude depending on the body. The longitude systems of most of those bodies with observable rigid surfaces have been defined by references to a surface feature such as a crater. The north pole is that pole of rotation that lies on the north side of the invariable plane of the solar system (near the ecliptic). The location of the Prime Meridian as well as the position of body's north pole on the celestial sphere may vary with time due to precession of the axis of rotation of the planet (or satellite). If the position angle of the body's Prime Meridian increases with time, the body has a direct (or prograde) rotation; otherwise the rotation is said to be retrograde.

In the absence of other information, the axis of rotation is assumed to be normal to the mean orbital plane; Mercury and most of the satellites are in this category. For many of the satellites, it is assumed that the rotation rate is equal to the mean orbital period. In the case of the giant planets, since their surface features are constantly changing and moving at various rates, the rotation of their magnetic fields is used as a reference instead. In the case of the Sun, even this criterion fails (because its magnetosphere is very complex and does not really rotate in a steady fashion), and an agreed-upon value for the rotation of its equator is used instead.

For *planetographic longitude*, west longitudes (i.e., longitudes measured positively to the west) are used when the rotation is prograde, and east longitudes (i.e., longitudes measured positively to the east) when the rotation is retrograde. In simpler terms, imagine a distant, non-orbiting observer viewing a planet as it rotates. Also suppose that this observer is within the plane of the planet's equator. A point on the Equator that passes directly in front of this observer later in time has a higher planetographic longitude than a point that did so earlier in time.

However, *planetocentric longitude* is always measured positively to the east, regardless of which way the planet rotates. *East* is defined as the counter-clockwise direction around the planet, as seen from above its north pole, and the north pole is whichever pole more closely aligns with the Earth's north pole. Longitudes traditionally have been written using "E" or "W" instead of "+" or "−" to indicate this polarity. For example, the following all mean the same thing:

- −91°
- 91°W
- +269°
- 269°E.

The reference surfaces for some planets (such as Earth and Mars) are ellipsoids of revolution for which the equatorial radius is larger than the polar radius; in other words, they are oblate spheroids. Smaller bodies (Io, Mimas, etc.) tend to be better approximated by triaxial ellipsoids; however, triaxial ellipsoids would render many computations more complicated, especially those related to map projections. Many projections would lose their elegant and popular properties. For this reason spherical reference surfaces are frequently used in mapping programs.

The modern standard for maps of Mars (since about 2002) is to use planetocentric coordinates. The meridian of Mars is located at Airy-0 crater.

Tidally-locked bodies have a natural reference longitude passing through the point nearest to their parent body: 0° the center of the primary-facing hemisphere, 90° the center of the leading hemisphere, 180° the center of the anti-primary hemisphere, and 270° the center of the trailing hemisphere. However, libration due to non-circular orbits or axial tilts causes this point to move around any fixed point on the celestial body like an analemma.

Latitude

In geography, latitude is a geographic coordinate that specifies the north–south position of a point on the Earth's surface. Latitude is an angle (defined below) which ranges from 0° at the Equator to 90° (North or South) at the poles. Lines of constant latitude, or parallels, run east–west as circles parallel to the equator. Latitude is used together with longitude to specify the precise location of features on the surface of the Earth.

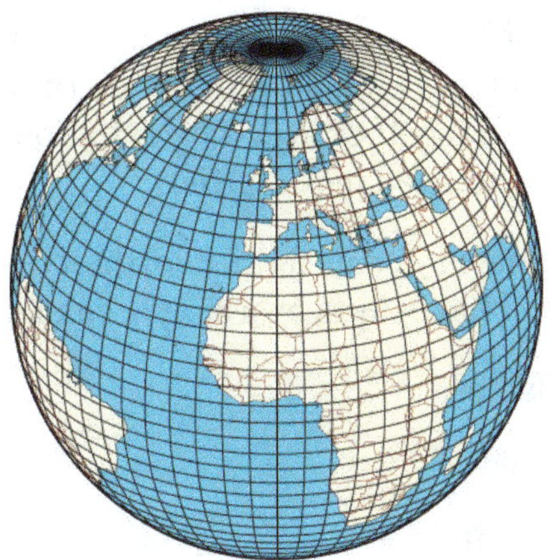

A graticule on the Earth as a sphere or an ellipsoid. The lines from pole to pole are lines of constant longitude, or meridians. The circles parallel to the equator are lines of constant latitude, or parallels. The graticule shows the latitude and longitude of points on the surface. In this example meridians are spaced at 6° intervals and parallels at 4° intervals.

Approximations Employed

Two levels of abstraction are employed in the definition of these coordinates. In the first step the physical surface is modelled by the geoid, a surface which approximates the mean sea level over the oceans and its continuation under the land masses. The second step is to approximate the geoid by a mathematically simpler reference surface. The simplest choice for the reference surface is a sphere, but the geoid is more accurately modelled by an ellipsoid. The definitions of latitude and longitude on such reference surfaces are detailed in the following sections. Lines of constant latitude and longitude together constitute a graticule on the reference surface. The latitude of a point on the *actual* surface is that of the corresponding point on the reference surface, the correspondence being along the normal to the reference surface which passes through the point on the physical surface. Latitude and longitude together with some specification of height constitute a geographic coordinate system as defined in the specification of the ISO 19111 standard.

Since there are many different reference ellipsoids, the precise latitude of a feature on the surface is not unique: this is stressed in the ISO standard which states that "without the full specification of the coordinate reference system, coordinates (that is latitude and longitude) are ambiguous at best and meaningless at worst". This is of great importance in accurate applications, such as a Global Positioning System (GPS), but in common usage, where high accuracy is not required, the reference ellipsoid is not usually stated.

In English texts the latitude angle, defined below, is usually denoted by the Greek lower-case letter phi (φ or). It is measured in degrees, minutes and seconds or decimal degrees, north or south of the equator.

The precise measurement of latitude requires an understanding of the gravitational field of the Earth, either to set up theodolites or to determine GPS satellite orbits. The study of the figure of the Earth together with its gravitational field is the science of geodesy. These topics are not discussed in this article.

This article relates to coordinate systems for the Earth: it may be extended to cover the Moon, planets and other celestial objects by a simple change of nomenclature.

The following lists are available:

- List of cities by latitude
- List of countries by latitude

History of Latitude Measurements

The Greeks studied the results of the measurements by the explorer Pytheas who voyaged to Britain and beyond, as far as the Arctic Circle (observing the midnight sun), in 325 BC. They used several methods to measure latitude, including the height of the Sun above the horizon at midday, measured using a gnōmōn (a word that originally meant an interpreter or judge); the length of the day at the summer solstice, and the elevation of the Sun at winter solstice.

The Greek Marinus of Tyre (AD 70–130) was the first to assign a latitude and longitude to every place on his maps.

From the late 9th century CE, the Arabian Kamal was used in equatorial regions, to measure the height of Polaris above the horizon. This instrument could only be used in latitudes close to the horizon.

The Mariner's astrolabe which gives the angle of the Sun from the horizon at noon, or the angle of a known star at night, was used from around the 15th to the 17th century. The observation of the Sun instead of Polaris enabled the measurement of latitude in the Southern hemisphere but required the use of solar declination tables. One of the most famous tables, but certainly not the first one, was published in 1496 by the Castilian Jew Abraham Zacut, then exiled in Portugal.

The Backstaff, which measures the length of a shadow, was used from the 16th century, but replaced by more accurate methods such as the Davis quadrant in the 16th century.

The sextant, which is still used to this day, was mentioned by Isaac Newton (1643–1727) in his unpublished writings, and first implemented about 1730 by John Hadley (1682–1744) and Thomas Godfrey (1704–1749).

Latitude on The Sphere

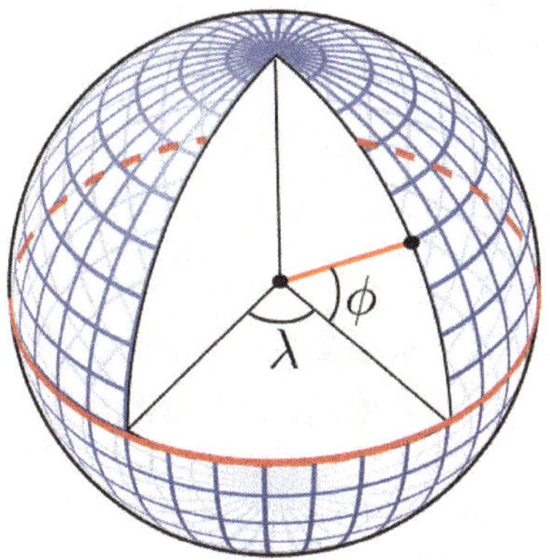

A perspective view of the Earth showing how latitude (φ) and longitude (λ) are defined on a spherical model. The graticule spacing is 10 degrees.

The Graticule on the Sphere

The graticule is formed by the lines of constant latitude and constant longitude, which are constructed with reference to the rotation axis of the Earth. The primary reference points are the poles where the axis of rotation of the Earth intersects the reference surface. Planes which contain the rotation axis intersect the surface at the meridians; and the angle between any one meridian plane and that through Greenwich (the Prime Meridian) defines the longitude: meridians are lines of constant longitude. The plane through the centre of the Earth and perpendicular to the rotation axis intersects the surface at a great circle called the Equator. Planes parallel to the equatorial plane intersect the surface in circles of constant latitude; these are the parallels. The Equator has a latitude of 0°, the North Pole has a latitude of 90° North (written 90° N or +90°), and the South Pole has a latitude of 90° South (written 90° S or −90°). The latitude of an arbitrary point is the angle between the equatorial plane and the radius to that point.

The latitude, as defined in this way for the sphere, is often termed the spherical latitude, to avoid ambiguity with auxiliary latitudes defined in subsequent sections of this article.

Named Latitudes on the Earth

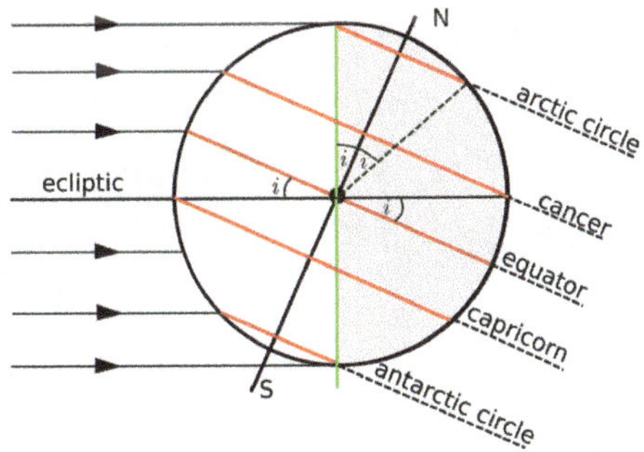

The orientation of the Earth at the December solstice.

Besides the equator, four other parallels are of significance:

Arctic Circle	66° 34′ (66.57°) N
Tropic of Cancer	23° 26′ (23.43°) N
Tropic of Capricorn	23° 26′ (23.43°) S
Antarctic Circle	66° 34′ (66.57°) S

The plane of the Earth's orbit about the Sun is called the ecliptic, and the plane perpendicular to the rotation axis of the Earth is the equatorial plane. The angle between the ecliptic and the equatorial plane is called variously the axial tilt, the obliquity, or the inclination of the ecliptic, and it is conventionally denoted by i. The latitude of the tropical circles is equal to i and the latitude of the polar circles is its complement ($90° - i$). The axis of rotation varies slowly over time and the values given here are those for the current epoch. The time variation is discussed more fully in the article on axial tilt.[b]

The figure shows the geometry of a cross-section of the plane perpendicular to the ecliptic and through the centres of the Earth and the Sun at the December solstice when the Sun is overhead at some point of the Tropic of Capricorn. The south polar latitudes below the Antarctic Circle are in daylight, whilst the north polar latitudes above the Arctic Circle are in night. The situation is reversed at the June solstice, when the Sun is overhead at the Tropic of Cancer. Only at latitudes in between the two tropics is it possible for the Sun to be directly overhead (at the zenith).

On map projections there is no universal rule as to how meridians and parallels should appear. The examples below show the named parallels (as red lines) on the commonly used Mercator projection and the Transverse Mercator projection. On the former the parallels are horizontal and the meridians are vertical, whereas on the latter there is no exact relationship of parallels and meridians with horizontal and vertical: both are complicated curves.

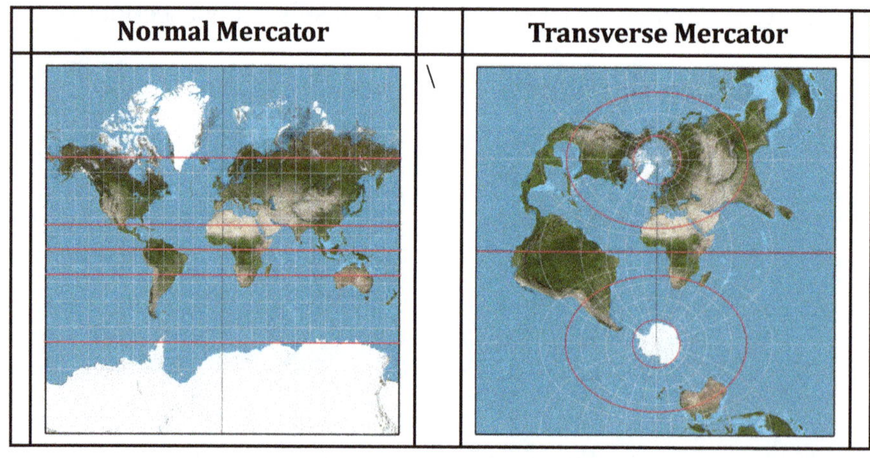

Meridian Distance on the Sphere

On the sphere the normal passes through the centre and the latitude (φ) is therefore equal to the angle subtended at the centre by the meridian arc from the equator to the point concerned. If the meridian distance is denoted by $m(\varphi)$ then

$$m(\varphi) = \frac{\pi}{180^\circ} R\varphi_{\text{degrees}} = R\varphi_{\text{radians}}$$

where R denotes the mean radius of the Earth. R is equal to 6,371 km or 3,959 miles. No higher accuracy is appropriate for R since higher-precision results necessitate an ellipsoid model. With this value for R the meridian length of 1 degree of latitude on the sphere is 111.2 km or 69.1 miles. The length of 1 minute of latitude is 1.853 km or 1.151 miles, used as the basis of the nautical mile.

Latitude on the Ellipsoid

Ellipsoids

In 1687 Isaac Newton published the *Philosophiæ Naturalis Principia Mathematica*, in which he proved that a rotating self-gravitating fluid body in equilibrium takes the form of an oblate ellipsoid. *Newton's result was confirmed by geodetic measurements in the 18th century. An oblate ellipsoid is the three-dimensional surface generated by the rotation of an ellipse about its shorter axis (minor axis). "Oblate ellipsoid of revolution" is abbreviated to 'ellipsoid' in the remainder of this article. (Ellipsoids which do not have an axis of symmetry are termed triaxial.)*

Many different reference ellipsoids have been used in the history of geodesy. In pre-satellite days they were devised to give a good fit to the geoid over the limited area of a survey but, with the advent of GPS, it has become natural to use reference ellipsoids (such as WGS84) with centre at the centre of mass of the Earth and minor axis aligned to the rotation axis of the Earth. These geocentric ellipsoids are usually within 100 m (330 ft) of the geoid. Since latitude is defined with respect to an ellipsoid, the position of a given point is different on each ellipsoid: one cannot exactly specify the latitude and longitude of a geographical feature without specifying the ellipsoid used. Many maps maintained by national agencies are based on older ellipsoids, so one must know how the latitude and longitude values are transformed from one ellipsoid to another. GPS handsets include software to carry out datum transformations which link WGS84 to the local reference ellipsoid with its associated grid.

The Geometry of the Ellipsoid

The shape of an ellipsoid of revolution is determined by the shape of the ellipse which is rotated about its minor (shorter) axis. Two parameters are required. One is invariably the equatorial radius, which is the semi-major axis, *a*. The other parameter is usually (1) the polar radius or semi-minor axis, *b*; or (2) the (first) flattening, *f*; or (3) the eccentricity, *e*. These parameters are not independent: they are related by

$$f = \frac{a-b}{a}, \qquad e^2 = 2f - f^2, \qquad b = a(1-f) = a\sqrt{1-e^2}.$$

Many other parameters appear in the study of geodesy, geophysics and map projections but they can all be expressed in terms of one or two members of the set *a*, *b*, *f* and *e*. Both *f* and *e* are small and often appear in series expansions in calculations; they are of the order 1/300 and 0.08 respectively. Values for a number of ellipsoids are given in Figure of the Earth. Reference ellipsoids are usually defined by the semi-major axis and the *inverse* flattening, 1/*f*. For example, the defining values for the WGS84 ellipsoid, used by all GPS devices, are

- *a* (equatorial radius): 6378137.0 m exactly
- 1/*f* (inverse flattening): 298.257223563 exactly

from which are derived

- *b* (polar radius): 6356752.3142 m
- e^2 (eccentricity squared): 0.00669437999014

The difference between the semi-major and semi-minor axes is about 21 km (13 miles) and as fraction of the semi-major axis it equals the flattening; on a computer the ellipsoid could be sized as 300 by 299 pixels. This would barely be distinguishable from a 300-by-300-pixel sphere, so illustrations usually exaggerate the flattening.

Geodetic and Geocentric Latitudes

The graticule on the ellipsoid is constructed in exactly the same way as on the sphere. The normal at a point on the surface of an ellipsoid does not pass through the centre, except for points on the equator or at the poles, but the definition of latitude remains unchanged as the angle between the normal and the equatorial plane. The terminology for latitude must be made more precise by distinguishing:

- Geodetic latitude: the angle between the normal and the equatorial plane. The standard notation in English publications is φ. This is the definition assumed when the word latitude is used without qualification. The definition must be accompanied with a specification of the ellipsoid.
- Geocentric latitude: the angle between the radius (from centre to the point on the surface) and the equatorial plane. (Figure below). There is no standard no-

tation: examples from various texts include ψ, q, φ', φ_c, φ_g. This article uses ψ.

- Spherical latitude: the angle between the normal to a spherical reference surface and the equatorial plane.
- Geographic latitude must be used with care. Some authors use it as a synonym for geodetic latitude whilst others use it as an alternative to the astronomical latitude.
- Latitude (unqualified) should normally refer to the geodetic latitude.

The importance of specifying the reference datum may be illustrated by a simple example. On the reference ellipsoid for WGS84, the centre of the Eiffel Tower has a geodetic latitude of 48° 51′ 29″ N, or 48.8583° N and longitude of 2° 17′ 40″ E or 2.2944°E. The same coordinates on the datum ED50 define a point on the ground which is 140 metres (460 feet) distant from the tower. A web search may produce several different values for the latitude of the tower; the reference ellipsoid is rarely specified.

Length of a Degree of Latitude

In Meridian arc and standard texts it is shown that the distance along a meridian from latitude φ to the equator is given by (φ in radians)

$$m(\varphi) = \int_0^{\varphi} M(\varphi')d\varphi' = a(1-e^2)\int_0^{\varphi}\left(1-e^2\sin^2\varphi'\right)^{-\frac{3}{2}}d\varphi'$$

where $M(\varphi)$ is the meridional radius of curvature.

The distance from the equator to the pole is

$$m_p = m\left(\frac{\pi}{2}\right)$$

For WGS84 this distance is 10001.965729 km.

The evaluation of the meridian distance integral is central to many studies in geodesy and map projection. It can be evaluated by expanding the integral by the binomial series and integrating term by term: see Meridian arc for details. The length of the meridian arc between two given latitudes is given by replacing the limits of the integral by the latitudes concerned. The length of a *small* meridian arc is given by

$$\delta m(\varphi) = M(\varphi)\delta\varphi = a(1-e^2)\left(1-e^2\sin^2\varphi\right)^{-\frac{3}{2}}\delta\varphi$$

φ	Δ1 lat	Δ1 long
0°	110.574 km	111.320 km
15°	110.649 km	107.550 km
30°	110.852 km	96.486 km
45°	111.132 km	78.847 km
60°	111.412 km	55.800 km
75°	111.618 km	28.902 km
90°	111.694 km	0.000 km

When the latitude difference is 1 degree, corresponding to π/180 radians, the arc distance is about

$$\Delta_{\text{lat}}^{1} = \frac{\pi a\left(1-e^{2}\right)}{180^{\circ}\left(1-e^{2}\sin^{2}\phi\right)^{\frac{3}{2}}}$$

The distance in metres (correct to 0.01 metre) between latitudes φ – 0.5 degrees and φ + 0.5 degrees on the WGS84 spheroid is

$$\Delta_{\text{lat}}^{1} = 111132.954 - 559.822\cos 2\varphi + 1.175\cos 4\varphi$$

The variation of this distance with latitude (on WGS84) is shown in the table along with the length of a degree of longitude (east-west distance):

$$\Delta_{\text{long}}^{1} = \frac{\pi a\cos\varphi}{180^{\circ}\sqrt{1-e^{2}\sin^{2}\varphi}}$$

A calculator for any latitude is provided by the U.S. Government's National Geospatial-Intelligence Agency (NGA).

Historically a nautical mile was defined as the length of one minute of arc along a meridian of a spherical earth. An ellipsoid model leads to a variation of the nautical mile with latitude. This was resolved by defining the nautical mile to be exactly 1,852 metres.

Auxiliary Latitudes

There are six auxiliary latitudes that have applications to special problems in geodesy, geophysics and the theory of map projections:

- Geocentric latitude
- Reduced (or parametric) latitude

- Rectifying latitude
- Authalic latitude
- Conformal latitude
- Isometric latitude

The definitions given in this section all relate to locations on the reference ellipsoid but the first two auxiliary latitudes, like the geodetic latitude, can be extended to define a three-dimensional geographic coordinate system. The remaining latitudes are not used in this way; they are used *only* as intermediate constructs in map projections of the reference ellipsoid to the plane or in calculations of geodesics on the ellipsoid. Their numerical values are not of interest. For example, no one would need to calculate the authalic latitude of the Eiffel Tower.

The expressions below give the auxiliary latitudes in terms of the geodetic latitude, the semi-major axis, a, and the eccentricity, e. The forms given are, apart from notational variants, those in the standard reference for map projections, namely "Map projections: a working manual" by J. P. Snyder. Derivations of these expressions may be found in Adams and online publications by Osborne and Rapp.

Geocentric Latitude

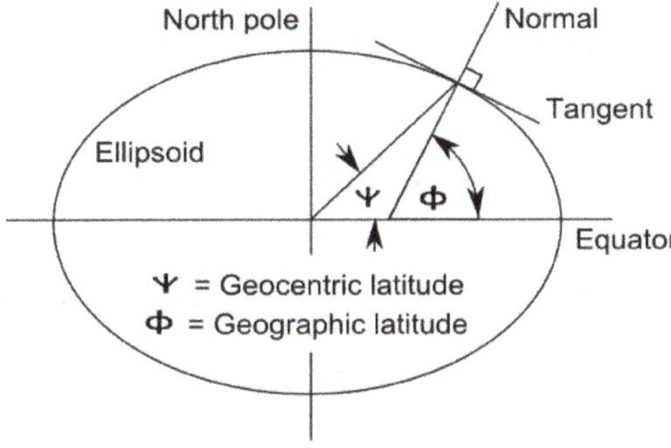

The definition of geodetic (or geographic) and geocentric latitudes.

The geocentric latitude is the angle between the equatorial plane and the radius from the centre to a point on the surface. The relation between the geocentric latitude (ψ) and the geodetic latitude (φ) is derived in the above references as

$$\psi(\varphi) = \tan^{-1}\left((1-e^2)\tan\varphi\right).$$

The geodetic and geocentric latitudes are equal at the equator and at the poles. The value of the squared eccentricity is approximately 0.0067 (depending on the choice of ellipsoid) and the maximum difference, $\varphi - \psi$, is about 11.5 minutes of arc at a geodetic latitude of $45° 5'$.

Reduced (or Parametric) Latitude

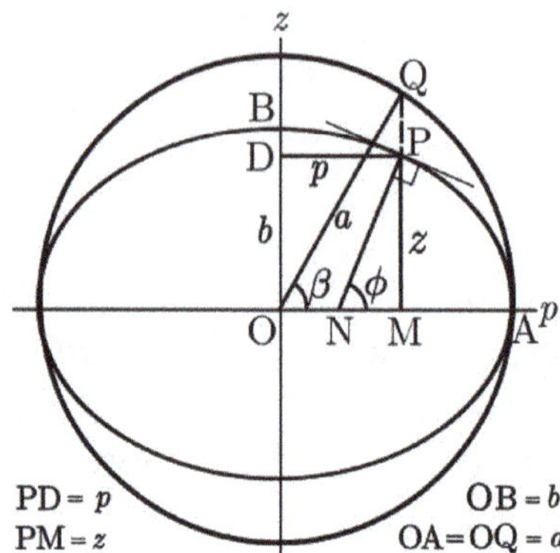

Definition of the reduced latitude (β) on the ellipsoid.

The reduced or parametric latitude, β, is defined by the radius drawn from the centre of the ellipsoid to that point Q on the surrounding sphere (of radius a) which is the projection parallel to the Earth's axis of a point P on the ellipsoid at latitude φ. It was introduced by Legendre and Bessel who solved problems for geodesics on the ellipsoid by transforming them to an equivalent problem for spherical geodesics by using this smaller latitude. Bessel's notation, $u(\varphi)$, is also used in the current literature. The reduced latitude is related to the geodetic latitude by:

$$\beta(\phi) = \tan^{-1}\left(\sqrt{1 - e^2}\tan\phi\right)$$

The alternative name arises from the parameterization of the equation of the ellipse describing a meridian section. In terms of Cartesian coordinates p, the distance from the minor axis, and z, the distance above the equatorial plane, the equation of the ellipse is:

$$\frac{p^2}{a^2} + \frac{z^2}{b^2} = 1.$$

The Cartesian coordinates of the point are parameterized by

$$p = a\cos\beta, \qquad z = b\sin\beta;$$

Cayley suggested the term *parametric latitude* because of the form of these equations.

The reduced latitude is not used in the theory of map projections. Its most important application is in the theory of ellipsoid geodesics. (Vincenty, Karney).

Rectifying Latitude

The rectifying latitude, μ, is the meridian distance scaled so that its value at the poles is equal to 90 degrees or π/2 radians:

$$\mu(\varphi) = \frac{\pi}{2}\frac{m(\varphi)}{m_{\mathrm{p}}}$$

where the meridian distance from the equator to a latitude φ is

$$m(\varphi) = a(1-e^2)\int_0^{\varphi}\left(1-e^2\sin^2\varphi'\right)^{-\frac{3}{2}}d\varphi',$$

and the length of the meridian quadrant from the equator to the pole (the polar distance) is

$$m_{\mathrm{p}} = m\left(\frac{\pi}{2}\right).$$

Using the rectifying latitude to define a latitude on a sphere of radius

$$R = \frac{2m_{\mathrm{p}}}{\pi}$$

defines a projection from the ellipsoid to the sphere such that all meridians have true length and uniform scale. The sphere may then be projected to the plane with an equirectangular projection to give a double projection from the ellipsoid to the plane such that all meridians have true length and uniform meridian scale. An example of the use of the rectifying latitude is the Equidistant conic projection. (Snyder, Section 16). The rectifying latitude is also of great importance in the construction of the Transverse Mercator projection.

Authalic Latitude

The authalic latitude, ξ, gives an area-preserving transformation to a sphere.

$$\xi(\varphi) = \sin^{-1}\left(\frac{q(\varphi)}{q_{\mathrm{p}}}\right)$$

where

$$\begin{aligned} q(\varphi) &= \frac{(1-e^2)\sin\varphi}{1-e^2\sin^2\varphi} - \frac{1-e^2}{2e}\ln\left(\frac{1-e\sin\varphi}{1+e\sin\varphi}\right) \\ &= \frac{(1-e^2)\sin\varphi}{1-e^2\sin^2\varphi} + \frac{1-e^2}{e}\tanh^{-1}(e\sin\varphi) \end{aligned}$$

and

$$q_{\rm p} = q\left(\frac{\pi}{2}\right) \quad = 1 - \frac{1-e^2}{2e}\ln\left(\frac{1-e}{1+e}\right) \quad = 1 + \frac{1-e^2}{e}\tanh^{-1} e$$

and the radius of the sphere is taken as

$$R_q = a\sqrt{\frac{q_{\rm p}}{2}}.$$

An example of the use of the authalic latitude is the Albers equal-area conic projection.

Conformal Latitude

where gd(x) is the Gudermannian function. The conformal latitude defines a transformation from the ellipsoid to a sphere of *arbitrary* radius such that the angle of intersection between any two lines on the ellipsoid is the same as the corresponding angle on the sphere (so that the shape of *small* elements is well preserved). A further conformal transformation from the sphere to the plane gives a conformal double projection from the ellipsoid to the plane. This is not the only way of generating such a conformal projection. For example, the 'exact' version of the Transverse Mercator projection on the ellipsoid is not a double projection. (It does, however, involve a generalisation of the conformal latitude to the complex plane).

$$\begin{aligned}
\chi(\phi) &= 2\tan^{-1}\left[\left(\frac{1+\sin\phi}{1-\sin\phi}\right)\left(\frac{1-\mathrm{e}\sin\phi}{1+\mathrm{e}\sin\phi}\right)^{\mathrm{e}}\right]^{\frac{1}{2}} - \frac{\pi}{2} \\
&= 2\tan^{-1}\left[\tan\left(\frac{\phi}{2}+\frac{\pi}{4}\right)\left(\frac{1-\mathrm{e}\sin\phi}{1+\mathrm{e}\sin\phi}\right)^{\frac{\mathrm{e}}{2}}\right] - \frac{\pi}{2} \\
&= \sin^{-1}\left[\tanh\left(\tanh^{-1}(\sin\phi) - \mathrm{e}\tanh^{-1}(\mathrm{e}\sin\phi)\right)\right] \\
&= \mathrm{gd}\left[\mathrm{gd}^{-1}(\phi) - \mathrm{e}\tanh^{-1}(\mathrm{e}\sin\phi)\right]
\end{aligned}$$

Isometric Latitude

The isometric latitude is conventionally denoted by ψ: it is used in the development of the ellipsoidal versions of the normal Mercator projection and the Transverse Mercator projection. The name "isometric" arises from the fact that at any point on the ellipsoid equal increments of ψ and longitude λ give rise to equal distance displacements along the meridians and parallels respectively. The graticule defined by the lines of constant ψ and constant λ, divides the surface of the ellipsoid into a mesh of squares (of varying size). The isometric latitude is zero at the equator but rapidly diverges from the geodetic latitude, tending to infinity at the poles. The conventional notation is given in Snyder :

$$\begin{aligned}\psi(\phi) &= \ln\left[\tan\left(\frac{\pi}{4}+\frac{\phi}{2}\right)\right]+\frac{e}{2}\ln\left[\frac{1-e\sin\phi}{1+e\sin\phi}\right]\\ &= \tanh^{-1}(\sin\phi)-e\tanh^{-1}(e\sin\phi)\\ &= \mathrm{gd}^{-1}(\phi)-e\tanh^{-1}(e\sin\phi).\end{aligned}$$

For the *normal* Mercator projection (on the ellipsoid) this function defines the spacing of the parallels: if the length of the equator on the projection is E (units of length or pixels) then the distance, y, of a parallel of latitude φ from the equator is

$$y(\varphi)=\frac{E}{2\pi}\psi(\varphi).$$

The isometric latitude ψ is closely related to the conformal latitude χ:

$$\psi(\varphi)=gd^{-1}\chi(\varphi).$$

Inverse Formulae and Series

The formulae in the previous sections give the auxiliary latitude in terms of the geodetic latitude. The expressions for the geocentric and reduced latitudes may be inverted directly but this is impossible in the four remaining cases: the rectifying, authalic, conformal, and isometric latitudes. There are two methods of proceeding. The first is a numerical inversion of the defining equation for each and every particular value of the auxiliary latitude. The methods available are fixed-point iteration and Newton–Raphson root finding. The other, more useful, approach is to express the auxiliary latitude as a series in terms of the geodetic latitude and then invert the series by the method of Lagrange reversion. Such series are presented by Adams who uses Taylor series expansions and gives coefficients in terms of the eccentricity. Osborne derives series to arbitrary order by using the computer algebra package Maxima and expresses the coefficients in terms of both eccentricity and flattening. The series method is not applicable to the isometric latitude and one must use the conformal latitude in an intermediate step.

Numerical Comparison of Auxiliary Latitudes

The following plot shows the magnitude of the difference between the geodetic latitude, (denoted as the "common" latitude on the plot), and the auxiliary latitudes other than the isometric latitude (which diverges to infinity at the poles). In every case the geodetic latitude is the greater. The differences shown on the plot are in arc minutes. The horizontal resolution of the plot fails to make clear that the maxima of the curves are not at 45° but calculation shows that they are within a few arc minutes of 45°. Some representative data points are given in the table following the plot. Note the closeness of the conformal and geocentric latitudes. This was exploited in the days of hand calculators to expedite the construction of map projections.

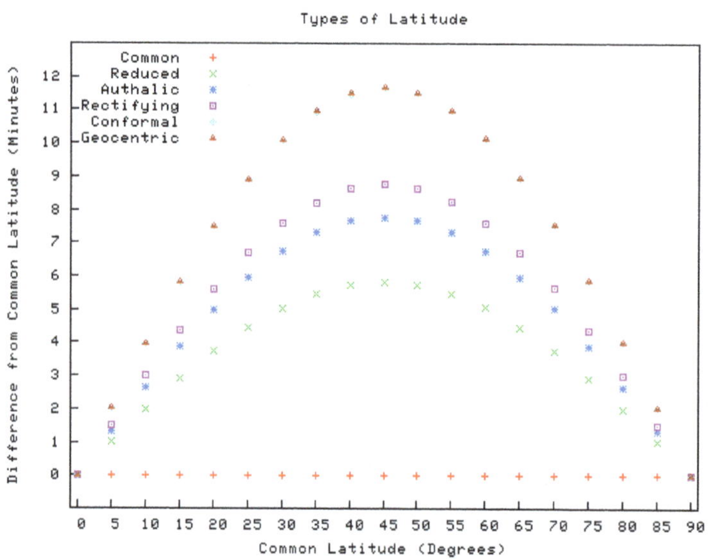

Approximate difference from geodetic latitude (φ)					
φ	**Reduced** $\varphi - \beta$	**Authalic** $\varphi - \xi$	**Rectifying** $\varphi - \mu$	**Conformal** $\varphi - \chi$	**Geocentric** $\varphi - \psi$
0°	0.00′	0.00′	0.00′	0.00′	0.00′
15°	2.91′	3.89′	4.37′	5.82′	5.82′
30°	5.05′	6.73′	7.57′	10.09′	10.09′
45°	5.84′	7.78′	8.76′	11.67′	11.67′
60°	5.06′	6.75′	7.59′	10.12′	10.13′
75°	2.92′	3.90′	4.39′	5.85′	5.85′
90°	0.00′	0.00′	0.00′	0.00′	0.00′

Latitude and Coordinate Systems

The geodetic latitude, or any of the auxiliary latitudes defined on the reference ellipsoid, constitutes with longitude a two-dimensional coordinate system on that ellipsoid. To define the position of an arbitrary point it is necessary to extend such a coordinate system into three dimensions. Three latitudes are used in this way: the geodetic, geocentric and reduced latitudes are used in geodetic coordinates, spherical polar coordinates and ellipsoidal coordinates respectively.

Geodetic Coordinates

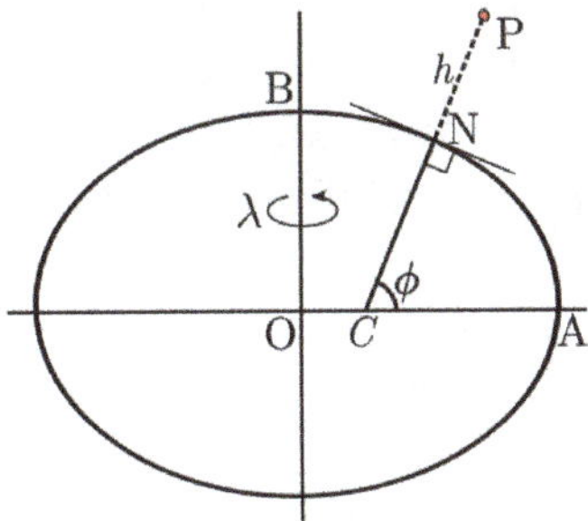

Geodetic coordinates $P(\phi,\lambda,h)$

At an arbitrary point P consider the line PN which is normal to the reference ellipsoid. The geodetic coordinates $P(\phi,\lambda,h)$ are the latitude and longitude of the point N on the ellipsoid and the distance PN. This height differs from the height above the geoid or a reference height such as that above mean sea level at a specified location. The direction of PN will also differ from the direction of a vertical plumb line. The relation of these different heights requires knowledge of the shape of the geoid and also the gravity field of the Earth.

Spherical Polar Coordinates

The geocentric latitude ψ is the complement of the polar angle θ in conventional spherical polar coordinates in which the coordinates of a point are $P(r,\theta,\lambda)$ where r is the distance of P from the centre O, θ is the angle between the radius vector and the polar axis and λ is longitude. Since the normal at a general point on the ellipsoid does not pass through the centre it is clear that points on the normal, which all have the same geodetic latitude, will have differing geocentric latitudes. Spherical polar coordinate systems are used in the analysis of the gravity field.

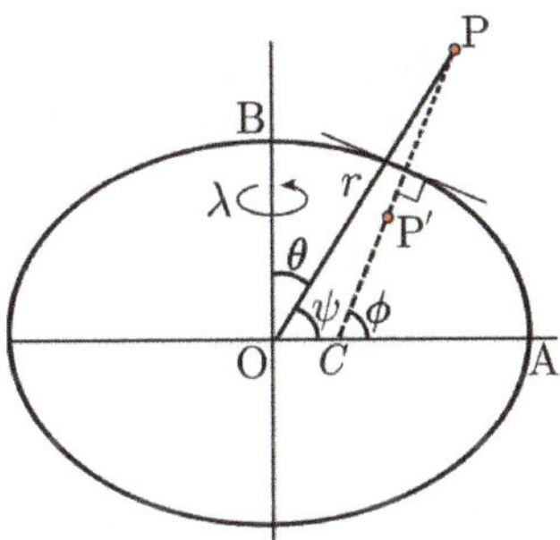

Geocentric coordinate related to spherical polar coordinates $P(r,\theta,\lambda)$

Ellipsoidal Coordinates

The reduced latitude can also be extended to a three-dimensional coordinate system. For a point P not on the reference ellipsoid (semi-axes OA and OB) construct an auxiliary ellipsoid which is confocal (same foci F, F′) with the reference ellipsoid:

the necessary condition is that the product *ae* of semi-major axis and eccentricity is the same for both ellipsoids. Let u be the semi-minor axis (OD) of the auxiliary ellipsoid. Further let β be the reduced latitude of P on the auxiliary ellipsoid. The set (u,β,λ) define the ellipsoid coordinates. These coordinates are the natural choice in models of the gravity field for a uniform distribution of mass bounded by the reference ellipsoid.

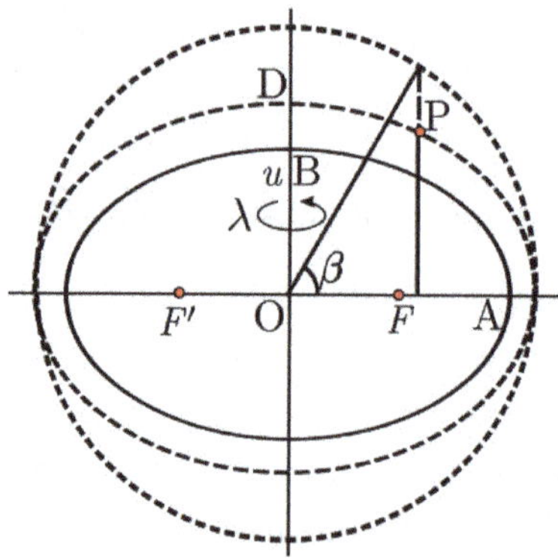

Ellipsoidal coordinates $P(u,\beta,\lambda)$

Coordinate Conversions

The relations between the above coordinate systems, and also Cartesian coordinates are not presented here. The transformation between geodetic and Cartesian coordinates may be found in Geographic coordinate conversion. The relation of Cartesian and spherical polars is given in Spherical coordinate system. The relation of Cartesian and ellipsoidal coordinates is discussed in Torge.

Astronomical Latitude

Astronomical latitude (Φ) is the angle between the equatorial plane and the true vertical at a point on the surface. The true vertical, the direction of a plumb line, is also the direction of the gravity acceleration, the resultant of the gravitational acceleration (mass-based) and the centrifugal acceleration at that latitude. Astronomic latitude is calculated from angles measured between the zenith and stars whose declination is accurately known.

In general the true vertical at a point on the surface does not exactly coincide with either the normal to the reference ellipsoid or the normal to the geoid. The angle between the astronomic and geodetic normals is usually a few seconds of arc but it is important in geodesy. The reason why it differs from the normal to the geoid is, because the geoid is an idealized, theoretical shape "at mean sea level". Points on the real surface of the earth are usually above or below this idealized geoid surface and here the true vertical can vary slightly. Also, the true vertical at a point at a specific time is influenced by tidal forces, which the theoretical geoid averages out.

The coordinate astronomers use in a similar way to specify the angular position of stars north/south of the celestial equator, nor with ecliptic latitude, the coordinate that astronomers use to specify the angular position of stars north/south of the ecliptic.

Geostationary Orbit

Two geostationary satellites in same orbit

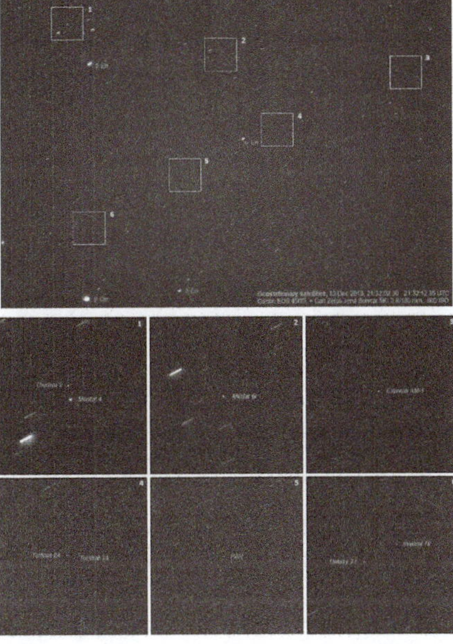

A 5 × 6 degree view of a part of the geostationary belt, showing several geostationary satellites. Those with inclination 0° form a diagonal belt across the image; a few objects with small inclinations to the equator are visible above this line. The satellites are pinpoint, while stars have created small trails due to the Earth's rotation.

A geostationary orbit, geostationary Earth orbit or geosynchronous equatorial orbit (GEO) is a circular orbit 35,786 kilometres (22,236 mi) above the Earth's equator and following the direction of the Earth's rotation. An object in such an orbit has an orbital period equal to the Earth's rotational period (one sidereal day) and thus appears motionless, at a fixed position in the sky, to ground observers. Communications satellites and weather satellites are often placed in geostationary orbits, so that the satellite antennas (located on Earth) that communicate with them do not have to rotate to track them, but can be pointed permanently at the position in the sky where the satellites are located. Using this characteristic, ocean color satellites with visible and near-infrared light sensors (e.g. the Geostationary Ocean Color Imager (GOCI)) can also be operated in geostationary orbit in order to monitor sensitive changes of ocean environments.

A geostationary orbit is a particular type of geosynchronous orbit, the distinction being that while an object in geosynchronous orbit returns to the same point in the sky at the same time each day, an object in geostationary orbit never leaves that position.

History

The notion of a geostationary space station equipped with radio communication was published in 1928 by Herman Potočnik. The first appearance of a geostationary orbit in popular literature was in the first Venus Equilateral story by George O. Smith, but Smith did not go into details. British science fiction author Arthur C. Clarke disseminated the idea widely, with more details on how it would work, in a 1945 paper entitled "Extra-Terrestrial Relays — Can Rocket Stations Give Worldwide Radio Coverage?", published in *Wireless World* magazine. Clarke acknowledged the connection in his introduction to *The Complete Venus Equilateral.* The orbit, which Clarke first described as useful for broadcast and relay communications satellites, is sometimes called the Clarke Orbit. Similarly, the Clarke Belt is the part of space about 35,786 km (22,236 mi) above sea level, in the plane of the equator, where near-geostationary orbits may be implemented. The Clarke Orbit is about 265,000 km (165,000 mi) in circumference.

Practical Uses

Most commercial communications satellites, broadcast satellites and SBAS satellites operate in geostationary orbits. A geostationary transfer orbit is used to move a satellite from low Earth orbit (LEO) into a geostationary orbit. The first satellite placed into a geostationary orbit was the Syncom-3, launched by a Delta D rocket in 1964.

A worldwide network of operational geostationary meteorological satellites is used to provide visible and infrared images of Earth's surface and atmosphere. These satellite systems include:

- the United States GOES
- Meteosat, launched by the European Space Agency and operated by the European Weather Satellite Organization, EUMETSAT

- the Japanese Himawari
- Chinese Fengyun
- India's INSAT series

A statite, a hypothetical satellite that uses a solar sail to modify its orbit, could theoretically hold itself in a geostationary "orbit" with different altitude and/or inclination from the "traditional" equatorial geostationary orbit.

Orbital Stability

A geostationary orbit can only be achieved at an altitude very close to 35,786 km (22,236 mi) and directly above the equator. This equates to an orbital velocity of 3.07 km/s (1.91 mi/s) and an orbital period of 1,436 minutes, which equates to almost exactly one sidereal day (23.934461223 hours). This ensures that the satellite will match the Earth's rotational period and has a stationary footprint on the ground. All geostationary satellites have to be located on this ring.

A combination of lunar gravity, solar gravity, and the flattening of the Earth at its poles causes a precession motion of the orbital plane of any geostationary object, with an orbital period of about 53 years and an initial inclination gradient of about 0.85° per year, achieving a maximal inclination of 15° after 26.5 years. To correct for this orbital perturbation, regular orbital stationkeeping manoeuvres are necessary, amounting to a delta-v of approximately 50 m/s per year.

A second effect to be taken into account is the longitude drift, caused by the asymmetry of the Earth – the equator is slightly elliptical. There are two stable (at 75.3°E and 104.7°W) and two unstable (at 165.3°E and 14.7°W) equilibrium points. Any geostationary object placed between the equilibrium points would (without any action) be slowly accelerated towards the stable equilibrium position, causing a periodic longitude variation. The correction of this effect requires station-keeping maneuvers with a maximal delta-v of about 2 m/s per year, depending on the desired longitude.

Solar wind and radiation pressure also exert small forces on satellites; over time, these cause them to slowly drift away from their prescribed orbits.

In the absence of servicing missions from the Earth or a renewable propulsion method, the consumption of thruster propellant for station keeping places a limitation on the lifetime of the satellite. Hall-effect thrusters, which are currently in use, have the potential to prolong the service life of a satellite by providing high-efficiency electric propulsion.

Communications

Satellites in geostationary orbits are far enough away from Earth that communication latency becomes significant — about a quarter of a second for a trip from one ground-based transmitter to the satellite and back to another ground-based transmitter; close to half a second for a round-trip communication from one Earth station to another and then back to the first.

For example, for ground stations at latitudes of $\varphi = \pm 45^\circ$ on the same meridian as the satellite, the time taken for a signal to pass from Earth to the satellite and back again can be computed using the cosine rule, given the geostationary orbital radius r (derived below), the Earth's radius R and the speed of light c, as

$$\Delta t = \frac{2}{c}\sqrt{R^2 + r^2 - 2Rr\cos\varphi} \approx 253\,ms.$$

(Note that r is the orbital radius, the distance from the centre of the Earth, not the height above the equator.)

This delay presents problems for latency-sensitive applications such as voice communication.

Geostationary satellites are directly overhead at the equator and become lower in the sky the further north or south one travels. As the observer's latitude increases, communication becomes more difficult due to factors such as atmospheric refraction, Earth's thermal emission, line-of-sight obstructions, and signal reflections from the ground or nearby structures. At latitudes above about 81°, geostationary satellites are below the horizon and cannot be seen at all. Because of this, some Russian communication satellites have used elliptical Molniya and Tundra orbits, which have excellent visibility at high latitudes.

Orbit Allocation

Satellites in geostationary orbit must all occupy a single ring above the equator. The requirement to space these satellites apart to avoid harmful radio-frequency interference during operations means that there are a limited number of orbital "slots" available, thus only a limited number of satellites can be operated in geostationary orbit. This has led to conflict between different countries wishing access to the same orbital slots (countries near the same longitude but differing latitudes) and radio frequencies. These disputes are addressed through the International Telecommunication Union's allocation mechanism. In the 1976 Bogotá Declaration, eight countries located on the Earth's equator claimed sovereignty over the geostationary orbits above their territory, but the claims gained no international recognition.

Limitations to Usable Life of Geostationary Satellites

When they run out of thruster fuel, the satellites are at the end of their service life, as they are no longer able to stay in their allocated orbital position. The transponders and other onboard systems generally outlive the thruster fuel and, by stopping N–S station keeping, some satellites can continue to be used in inclined orbits (where the orbital track appears to follow a figure-eight loop centred on the equator), or else be elevated to a "graveyard" disposal orbit.

Derivation of Geostationary Altitude

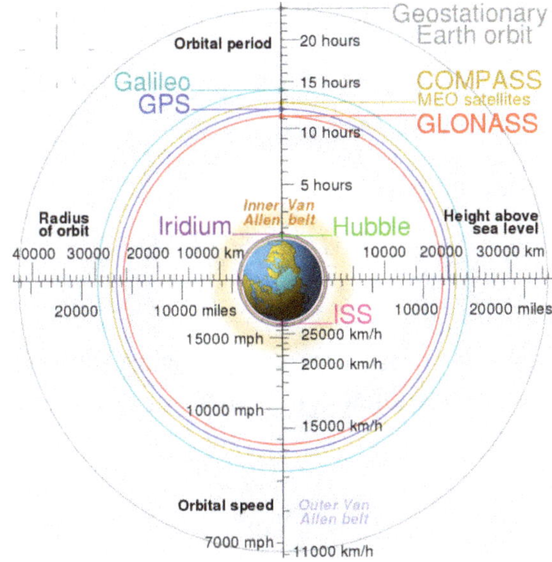

Comparison of geostationary Earth orbit with GPS, GLONASS, Galileo and Compass (medium Earth orbit) satellite navigation system orbits with the International Space Station, Hubble Space Telescope and Iridium constellation orbits, and the nominal size of the Earth. The Moon's orbit is around 9 times larger (in radius and length) than geostationary orbit.

In any circular orbit, the centripetal force required to maintain the orbit (F_c) is provided by the gravitational force on the satellite (F_g). To calculate the geostationary orbit altitude, one begins with this equivalence:

$$\mathbf{F}_c = \mathbf{F}_g.$$

By Newton's second law of motion, we can replace the forces F with the mass m of the object multiplied by the acceleration felt by the object due to that force:

$$m\mathbf{a}_c = m\mathbf{g}.$$

We note that the mass of the satellite m appears on both sides — geostationary orbit is independent of the mass of the satellite.[c] So calculating the altitude simplifies into calculating the point where the magnitudes of the centripetal acceleration required for orbital motion and the gravitational acceleration provided by Earth's gravity are equal.

The centripetal acceleration's magnitude is:

$$|\mathbf{a}_c| = \omega^2 r,$$

where ω is the angular speed, and r is the orbital radius as measured from the Earth's center of mass.

The magnitude of the gravitational acceleration is:

$$|\mathbf{g}| = \frac{GM}{r^2},$$

where M is the mass of Earth, 5.9736×10^{24} kg, and G is the gravitational constant, $(6.67428 \pm 0.00067) \times 10^{-11}$ m^3 kg^{-1} s^{-2}.

Equating the two accelerations gives:

$$r^3 = \frac{GM}{\omega^2} \rightarrow r = \sqrt[3]{\frac{GM}{\omega^2}}.$$

The product GM is known with much greater precision than either factor alone; it is known as the geocentric gravitational constant $\mu = 398{,}600.4418 \pm 0.0008$ km^3 s^{-2}. Hence

$$r = \sqrt[3]{\frac{\mu}{\omega^2}}$$

The angular speed ω is found by dividing the angle travelled in one revolution (360° = 2π rad) by the orbital period (the time it takes to make one full revolution). In the case of a geostationary orbit, the orbital period is one sidereal day, or 86164.09054 s). This gives

$$\omega \approx \frac{2\pi \text{ rad}}{86164 \text{ s}} \approx 7.2921 \times 10^{-5} \text{ rad/s}.$$

The resulting orbital radius is 42,164 kilometres (26,199 mi). Subtracting the Earth's equatorial radius, 6,378 kilometres (3,963 mi), gives the altitude of 35,786 kilometres (22,236 mi).

Orbital speed is calculated by multiplying the angular speed by the orbital radius:

$$v = \omega r \approx 3.0746 \text{km/s} \approx 11068 \text{km/h} \approx 6877.8 \text{mph}.$$

By the same formula, we can find the geostationary-type orbit of an object in relation to Mars (this type of orbit above is referred to as an areostationary orbit if it is above Mars). The geocentric gravitational constant GM (which is μ) for Mars has the value of 42,828 km^{3}s^{-2}, and the known rotational period (T) of Mars is 88,642.66 seconds. Since $\omega = 2\pi/T$, using the formula above, the value of ω is found to be approx 7.088218×10^{-5} s^{-1}. Thus $r^3 = 8.5243 \times 10^{12}$ km^3, whose cube root is 20,427 km; subtracting the equatorial radius of Mars (3396.2 km), we have 17,031 km.

Geodetic Datum

A geodetic datum or geodetic system is a coordinate system, and a set of reference points, used to locate places on the Earth (or similar objects). An approximate definition of sea level is the datum WGS 84, an ellipsoid, whereas a more accurate definition is Earth Gravitational Model 2008 (EGM2008), using at least 2,159 spherical harmonics. Other datums are defined for other areas or at other times; ED50 was defined in 1950 over Europe and differs from WGS 84 by a few hundred meters depending on where in Europe you look. Mars has no oceans and so no sea level, but at least two martian datums have been used to locate places there.

Datums are used in geodesy, navigation, and surveying by cartographers and satellite navigation systems to translate positions indicated on maps (paper or digital) to their real position on Earth. Each starts with an ellipsoid (stretched sphere), and then defines latitude, longitude and altitude coordinates. One or more locations on the Earth's surface is chosen as an anchor "base-point".

The difference in co-ordinates between datums is commonly referred to as *datum shift*. The datum shift between two particular datums can vary from one place to another within one country or region, and can be anything from zero to hundreds of meters (or several kilometers for some remote islands). The North Pole, South Pole and Equator will be in different positions on different datums, so True North will be slightly different. Different datums use different interpolations for the precise shape and size of the Earth (reference ellipsoids).

Because the Earth is an imperfect ellipsoid, localised datums can give a more accurate representation of the area of coverage than WGS 84. OSGB36, for example, is a better approximation to the geoid covering the British Isles than the global WGS 84 ellipsoid. However, as the benefits of a global system outweigh the greater accuracy, the global WGS 84 datum is becoming increasingly adopted.

Horizontal datums are used for describing a point on the Earth's surface, in latitude and longitude or another coordinate system. Vertical datums measure elevations or depths.

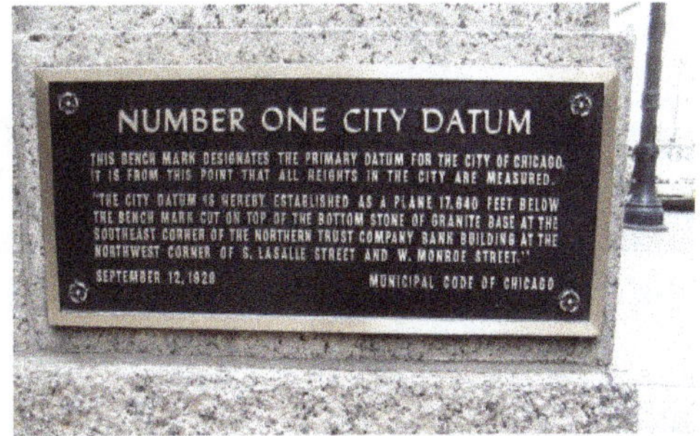

City of Chicago Datum Benchmark

Definition

In surveying and geodesy, a *datum* is a reference system or an approximation of the Earth's surface against which positional measurements are made for computing locations. Horizontal datums are used for describing a point on the Earth's surface, in latitude and longitude or another coordinate system. Vertical datums are used to measure elevations or underwater depths.

Horizontal Datum

The horizontal datum is the model used to measure positions on the Earth. A specific point on the Earth can have substantially different coordinates, depending on the datum used to make the measurement. There are hundreds of local horizontal datums around the world, usually referenced to some convenient local reference point. Contemporary datums, based on increasingly accurate measurements of the shape of the Earth, are intended to cover larger areas. The WGS 84 datum, which is almost identical to the NAD83 datum used in North America and the ETRS89 datum used in Europe, is a common standard datum.

For example, in Sydney there is a 200 metres (700 feet) difference between GPS coordinates configured in GDA (based on global standard WGS 84) and AGD (used for most local maps), which is an unacceptably large error for some applications, such as surveying or site location for scuba diving.

Vertical Datum

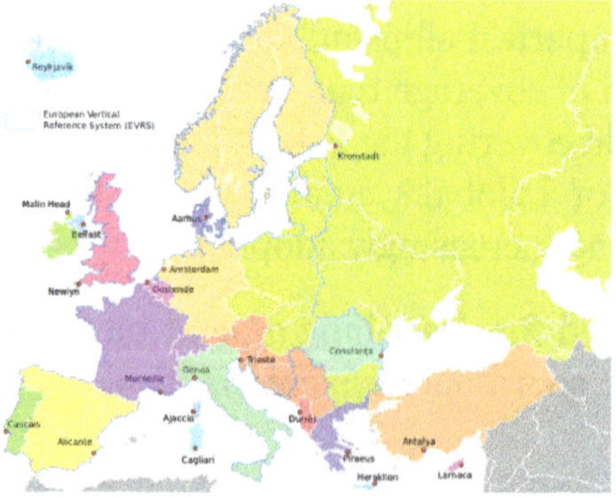

Vertical datums in Europe

A vertical datum is used as a reference point for elevations of surfaces and features on the Earth including terrain, bathymetry, water levels, and man-made structures. Vertical datums are either: tidal, based on sea levels; gravimetric, based on a geoid; or geodetic, based on the same ellipsoid models of the Earth used for computing horizontal datums.

In common usage, elevations are often cited in height above sea level, although what "sea level" actually means is a more complex issue than might at first be thought: the height of the sea surface at any one place and time is a result of numerous effects, including waves, wind and currents, atmospheric pressure, tides, topography, and even differences in the strength of gravity due to the presence of mountains etc.

For the purpose of measuring the height of objects on land, the usual datum used is mean sea level (MSL). This is a tidal datum which is described as the arithmetic mean of the hourly water elevation taken over a specific 19 years cycle. This definition averages out tidal highs and lows (caused by the gravitational effects of the sun and the moon) and short term variations. It will not remove the effects of local gravity strength, and so the height of MSL, relative to a geodetic datum, will vary around the world, and even around one country. Countries tend to choose the mean sea level at one specific point to be used as the standard "sea level" for all mapping and surveying in that country. (For example, in Great Britain, the national vertical datum, Ordnance Datum Newlyn, is based on what was mean sea level at Newlyn in Cornwall between 1915 and 1921). However, zero elevation as defined by one country is not the same as zero elevation defined by another (because MSL is not the same everywhere), which is why locally defined vertical datums differ from one another.

A different principle is used when choosing a datum for nautical charts. For safety reasons, a mariner must be able to know the minimum depth of water that could occur at any point. For this reason, depths and tides on a nautical chart are measured relative to chart datum, which is defined to be a level below which tide rarely falls. Exactly how this is chosen depends on the tidal regime in the area being charted and on the policy of the hydrographic office producing the chart in question; a typical definition is Lowest Astronomical Tide (the lowest tide predictable from the effects of gravity), or Mean Lower Low Water (the average lowest tide of each day), although MSL is sometimes used in waters with very low tidal ranges.

Conversely, if a ship is to safely pass under a low bridge or overhead power cable, the mariner must know the minimum clearance between the masthead and the obstruction, which will occur at high tide. Consequently, bridge clearances etc. are given relative to a datum based on high tide, such as Highest Astronomical Tide or Mean High Water Springs.

Sea level does not remain constant throughout geological time, and so tidal datums are less useful when studying very long-term processes. In some situations sea level does not apply at all — for instance for mapping Mars' surface — forcing the use of a different "zero elevation", such as mean radius.

A geodetic vertical datum takes some specific zero point, and computes elevations based on the geodetic model being used, without further reference to sea levels. Usually, the starting reference point is a tide gauge, so at that point the geodetic and tidal datums might match, but due to sea level variations, the two scales may not match elsewhere. An example of a gravity-based geodetic datum is NAVD88, used in North America, which is referenced to a point in Quebec, Canada. Ellipsoid-based datums such as WGS 84, GRS80 or NAD83 use a theoretical surface that may differ significantly from the geoid.

Geodetic Coordinates

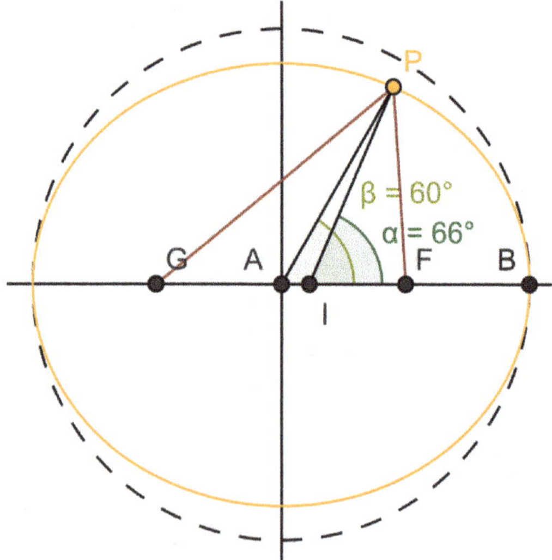

The same position on a spheroid has a different angle for latitude depending on whether the angle is measured from the normal line segment IP of the ellipsoid (angle α) or the line segment AP from the center (angle β). Note that the "flatness" of the spheroid (orange) in the image is greater than that of the Earth; as a result, the corresponding difference between the "geodetic" and "geocentric"latitudes is also exaggerated.

In geodetic coordinates the Earth's surface is approximated by an ellipsoid and locations near the surface are described in terms of latitude (ϕ), longitude (λ) and height (h).

Geodetic Versus Geocentric Latitude

It is important to note that geodetic latitude (Φ) (resp. altitude) is different from geocentric latitude (Φ') (resp. altitude). Geodetic latitude is determined by the angle between the equatorial plane and normal to the ellipsoid, whereas geocentric latitude is determined by the angle between the equatorial plane and line joining the point to the centre of the ellipsoid. Unless otherwise specified latitude is geodetic latitude.

Earth Reference Ellipsoid

Defining and Derived Parameters

The ellipsoid is completely parameterised by the semi-major axis a and the flattening f.

Parameter	Symbol
Semi-major axis	a
Reciprocal of flattening	$1/f$

From a and f it is possible to derive the semi-minor axis b, first eccentricity e and second eccentricity e' of the ellipsoid

Parameter	Value
semi-minor axis	$b = a(1 - f)$
First eccentricity squared	$e^2 = 1 - b^2/a^2 = 2f - f^2$
Second eccentricity squared	$e'^2 = a^2/b^2 - 1 = f(2 - f)/(1 - f)^2$

Parameters for Some Geodetic Systems

Australian Geodetic Datum 1966 [AGD66] and Australian Geodetic Datum 1984 (AGD84)

AGD66 and AGD84 both use the parameters defined by Australian National Spheroid

Australian National Spheroid (ANS)

ANS Defining Parameters		
Parameter	**Notation**	**Value**
semi-major axis	a	6 378 160.000 m
Reciprocal of Flattening	$1/f$	298.25

Geocentric Datum of Australia 1994 (GDA94)

GDA94 uses the parameters defined by GRS80

Geodetic Reference System 1980 (GRS80)

GRS80 Parameters		
Parameter	**Notation**	**Value**
semi-major axis	a	6 378 137 m
Reciprocal of flattening	$1/f$	298.257 222 101

see GDA Technical Manual document for more details; the value given above for the flattening is not exact.

World Geodetic System 1984 (WGS 84)

The Global Positioning System (GPS) uses the World Geodetic System 1984 (WGS 84) to determine the location of a point near the surface of the Earth.

WGS 84 Defining Parameters		
Parameter	**Notation**	**Value**
semi-major axis	**a**	**6 378 137.0 m**
Reciprocal of flattening	**$1/f$**	**298.257 223 563**

WGS 84 derived geometric constants		
Constant	**Notation**	**Value**
Semi-minor axis	b	6 356 752.3142 m
First eccentricity squared	e^2	6.694 379 990 14x10^{-3}
Second eccentricity squared	e'^2	6.739 496 742 28x10^{-3}

A more comprehensive list of geodetic systems can be found here

Conversion Calculations

Datum conversion is the process of converting the coordinates of a point from one datum system to another. Datum conversion may frequently be accompanied by a change of grid projection.

Reference

A reference datum is a known and constant surface which is used to describe the location of unknown points on the Earth. Since reference datums can have different radii and different center points, a specific point on the Earth can have substantially different coordinates depending on the datum used to make the measurement. There are hundreds of locally developed reference datums around the world, usually referenced to some convenient local reference point. Contemporary datums, based on increasingly accurate measurements of the shape of the Earth, are intended to cover larger areas.

The most common reference Datums in use in North America are NAD27, NAD83, and WGS 84.

The North American Datum of 1927 (NAD 27) is "the horizontal control datum for the United States that was defined by a location and azimuth on the Clarke spheroid of 1866, with origin at (the survey station) Meades Ranch (Kansas)." ... The geoidal height at Meades Ranch was assumed to be zero. "Geodetic positions on the North American Datum of 1927 were derived from the (coordinates of and an azimuth at Meades Ranch) through a readjustment of the triangulation of the entire network in which Laplace azimuths were introduced, and the Bowie method was used." (http://www.ngs.noaa.gov/faq.shtml#WhatDatum) NAD27 is a local referencing system covering North America.

The North American Datum of 1983 (NAD 83) is "The horizontal control datum for the United States, Canada, Mexico, and Central America, based on a geocentric origin and the Geodetic Reference System 1980 (GRS80). "This datum, designated as NAD 83 ...is based on the adjustment of 250,000 points including 600 satellite Doppler stations which constrain the system to a geocentric origin." NAD83 may be considered a local referencing system.

WGS 84 is the World Geodetic System of 1984. It is the reference frame used by the U.S. Department of Defense (DoD) and is defined by the National Geospatial-Intelligence Agency (NGA) (formerly the Defense Mapping Agency, then the National Imagery and Mapping Agency). WGS 84 is used by DoD for all its mapping, charting, surveying, and navigation needs, including its GPS "broadcast" and "precise" orbits. WGS 84 was defined in January 1987 using Doppler satellite surveying techniques. It was used as the reference frame for broadcast GPS Ephemerides (orbits) beginning January 23, 1987. At 0000 GMT January 2, 1994, WGS 84 was upgraded in accuracy using GPS measurements. The formal name then became WGS 84 (G730), since the upgrade date coincided with the start of GPS Week 730. It became the reference frame for broadcast orbits on June 28, 1994. At 0000 GMT September 30, 1996 (the start of GPS Week 873), WGS 84 was redefined again and was more closely aligned with International Earth Rotation Service (IERS) frame ITRF 94. It was then formally called WGS 84 (G873). WGS 84 (G873) was adopted as the reference frame for broadcast orbits on January 29, 1997. Another update brought it to WGS84(G1674).

The WGS 84 datum, within two meters of the NAD83 datum used in North America, is the only world referencing system in place today. WGS 84 is the default standard datum for coordinates stored in recreational and commercial GPS units.

Users of GPS are cautioned that they must always check the datum of the maps they are using. To correctly enter, display, and to store map related map coordinates, the datum of the map must be entered into the GPS map datum field.

Meridian Arc

In geodesy, a meridian arc measurement is the distance between two points with the same longitude, i.e., a segment of a meridian curve or its length. Two or more such determinations at different locations then specify the shape of the reference ellipsoid which best approximates the shape of the geoid. This process is called the determination of the Figure of the Earth. The earliest determinations of the size of a spherical Earth required a single arc. The latest determinations use astro-geodetic measurements and the methods of satellite geodesy to determine the reference ellipsoids.

Those interested in accurate expressions of the meridian arc for the WGS84 ellipsoid should consult the subsection entitled numerical expressions.

The Earth as a Sphere

Early estimations of Earth's size are recorded from Greece in the 4th century BC, and from scholars at the caliph's House of Wisdom in the 9th century. The first realistic value was calculated by Alexandrian scientist Eratosthenes about 240 BC. He knew that on the summer solstice at local noon the sun goes through the zenith in the ancient Egyptian city of Syene (Assuan). He also knew from his own measurements that, at the same moment in his hometown of Alexandria, the zenith distance was 1/50 of a full circle (7.2°).

Assuming that Alexandria was due north of Syene, Eratosthenes concluded that the distance between Alexandria and Syene must be 1/50 of Earth's circumference. Using data from caravan travels, he estimated the distance to be 5,000 stadia (about 500 nautical miles)—which implies a circumference of 252,000 stadia. Assuming the Attic stadion (185 m) this corresponds to 46,620 km, or 16% too great. However, if Eratosthenes used the Egyptian stadion (157.5 m) his measurement turns out to be 39,690 km, an error of only 1%. Syene is not precisely on the Tropic of Cancer and not directly south of Alexandria. The sun appears as a disk of 0.5°, and an estimate of the overland distance traveling along the Nile or through the desert couldn't be more accurate than about 10%.

Eratosthenes' estimation of Earth's size was accepted for nearly two thousand years. A similar method was used by Posidonius about 150 years later, and slightly better results were calculated in 827 by the grade measurement of the Caliph Al-Ma'mun.

The Earth as an Ellipsoid

Early literature uses the term *oblate spheroid* to describe a sphere "squashed at the poles". Modern literature uses the term "ellipsoid of revolution" in place of spheroid, although the qualifying words "of revolution" are usually dropped. An ellipsoid which is not an ellipsoid of revolution is called a triaxial ellipsoid. Spheroid and ellipsoid are used interchangeably in this article, with oblate implied if not stated.

The Eighteenth Century

In 1687 Newton had published in the Principia a proof that the earth was an oblate spheroid of flattening equal to 1/230. This was disputed by some, but not all, French scientists. A meridian arc of Picard was extended to a longer arc by J. D. Cassini over the period 1684–1718. The arc was measured with at least three latitude determinations, so they were able to deduce mean curvatures for the northern and southern halves of the arc, allowing a determination of the overall shape. The results indicated that the Earth was a *prolate* spheroid (with an equatorial radius less than the polar radius). (The history of the meridian arc from 1600 to 1880 is fully covered in the first chapter of *Geodesy* by Alexander Ross Clarke.) To resolve the issue, the French Academy of Sciences (1735) proposed expeditions to Peru (Bouguer, Louis Godin, de La Condamine, Antonio de Ulloa, Jorge Juan) and Lapland (Maupertuis, Clairaut, Camus, Le Monnier, Abbe Outhier, Celsius). The expedition to Peru is described in the French Geodesic Mission article and that to Lapland is described in the Torne Valley article. The resulting measurements at equatorial and polar latitudes confirmed that the earth was best modelled by an oblate spheroid, supporting Newton.

By the end of the century Delambre had remeasured and extended the French arc from Dunkirk to the Mediterranean. It was divided into five parts by four intermediate determinations of latitude. By combining the measurements together with those for the arc of Peru, ellipsoid shape parameters were determined and the distance between the equator and pole along the Paris Meridian was calculated as 5130762 toises as specified by the standard toise bar in Paris. Defining this distance as exactly 10000000 m led to the construction of a new standard metre bar as 0.5130762 toises.

The Nineteenth and Twentieth Centuries

In the 19th century, many astronomers and geodesists were engaged in detailed studies of the Earth's curvature along different meridian arcs. The analyses resulted in a great many model ellipsoids such as Plessis 1817, Airy 1830, Bessel 1830, Everest 1830, and Clarke 1866. A comprehensive list of ellipsoids is given under Earth ellipsoid.

Meridian Distance on the Ellipsoid

The determination of the meridian distance, that is the distance from the equator to a point at a latitude φ on the ellipsoid is an important problem in the theory of map projections, particularly the Transverse Mercator projection. Ellipsoids are normally specified in terms of the parameters defined above, a, b, f, but in theoretical work it is useful to define extra parameters, particularly the eccentricity, e, and the third flattening n. Only two of these parameters are independent and there are many relations between them:

$$f = \frac{a-b}{a}$$
$$b = a(1-f) = a\sqrt{1-e^2}$$

The meridian radius of curvature can be shown to be equal to

$$M(\varphi) = \frac{a(1-e^2)}{\left(1-e^2\sin^2\varphi\right)^{\frac{3}{2}}},$$

so that the arc length of an infinitesimal element of the meridian is $dm = M(\varphi)\, d\varphi$ (with φ in radians). Therefore, the meridian distance from the equator to latitude φ is

$$\mathrm{m}(\varphi) = \int_0^{\varphi} \mathrm{M}(\varphi)\mathrm{d}\varphi$$
$$= \mathrm{a}(1-\mathrm{e}^2)\int_0^{\varphi}\left(1-\mathrm{e}^2\sin^2\varphi\right)^{-\frac{3}{2}}\mathrm{d}\varphi.$$

The distance formula is simpler when written in terms of the parametric latitude,

$$m(\varphi) = b\int_0^{\beta}\sqrt{1+e'^2\sin^2\beta}\,d\beta,$$

where $\tan\beta = (1-f)\tan\varphi$ and $e'^2 = e^{2/1-}e^2$.

The distance from the equator to the pole, the quarter meridian, is

$$m_{\mathrm{p}} = m\left(\frac{\pi}{2}\right)$$

Even though latitude is normally confined to the range $[-\pi/2,\pi/2]$, all the formulae given here apply to measuring distance around the complete meridian ellipse (including the anti-meridian). Thus the ranges of φ, β, and the rectifying latitude μ, are unrestricted.

Relation to Elliptic Integrals

The above integral is related to a special case of an incomplete elliptic integral of the third kind. In the notation of the online NIST handbook.

$$m(\varphi) = a\left(1-e^2\right)\Pi(\varphi,\mathrm{e}^2,\mathrm{e})$$

It may also be written in terms of incomplete elliptic integrals of the second kind,

$$\begin{aligned} m(\varphi) &= a\left(E(\varphi,e) - \frac{e^2 \sin\varphi\cos\varphi}{\sqrt{1-e^2\sin^2\varphi}}\right) \\ &= a\left(E(\varphi,e) + \frac{d^2}{d\varphi^2}E(\varphi,e)\right) \\ &= bE(\beta, ie'). \end{aligned}$$

The quarter meridian can be expressed in terms of the complete elliptic integral of the second kind,

$$m_p = aE(e) = bE(ie').$$

The calculation (to arbitrary precision) of the elliptic integrals and approximations are also discussed in the NIST handbook. These functions are also implemented in computer algebra programs such as Mathematica and Maxima.

The Inverse Meridian Problem for the Ellipsoid

In some problems, we need to be able to solve the inverse problem: given m, determine φ. This may be solved by Newton's method, iterating

$$\varphi_{i+1} = \varphi_i - \frac{m(\varphi_i) - m}{M(\varphi_i)},$$

until convergence. A suitable starting guess is given by $\varphi_o = \mu$ where

$$\mu = \frac{\pi}{2}\frac{m}{m_p}$$

is the rectifying latitude. Note that it there is no need to differentiate the series for $m(\varphi)$, since the formula for the meridian radius of curvature $M(\varphi)$ can be used instead.

Alternatively, Helmert's series for the meridian distance can be reverted to give

$$\varphi = \mu + H_2' \sin 2\mu + H_4' \sin 4\mu + H_6' \sin 6\mu + H_8' \sin 8\mu + \cdots$$

where

$$H_2' = \tfrac{3}{2}n - \tfrac{27}{32}n^3 + \cdots, H_6' = \tfrac{151}{96}n^3 + \cdots, H_4' = \tfrac{21}{16}n^2 - \tfrac{55}{32}n^4 + \cdots, H_8' = \tfrac{1097}{512}n^4$$

Similarly, Bessel's series for m in terms of β can be reverted to give

$$\beta = \mu + B_2' \sin 2\mu + B_4' \sin 4\mu + B_6' \sin 6\mu + B_8' \sin 8\mu + \cdots$$

where

$$B'_2 = \frac{1}{2}n - \frac{9}{32}n^3 + \cdots, \qquad B'_6 = \frac{29}{96}n^3 - \cdots,$$

$$B'_4 = \frac{5}{16}n^2 - \frac{37}{96}n^4 + \cdots, \qquad B'_8 = \frac{539}{1536}n^4 - \cdots.$$

Legendre showed that the distance along a geodesic on an spheroid is the same as the distance along the perimeter of an ellipse. For this reason, the expression for m in terms of β and its inverse given above play a key role in the solution of the geodesic problem with m replaced by s, the distance along the geodesic, and β replaced by σ, the arc length on the auxiliary sphere. The requisite series extended to sixth order are given by Karney, Eqs. (17) & (21), with ε playing the role of n and τ playing the role of μ.

Axes Conventions

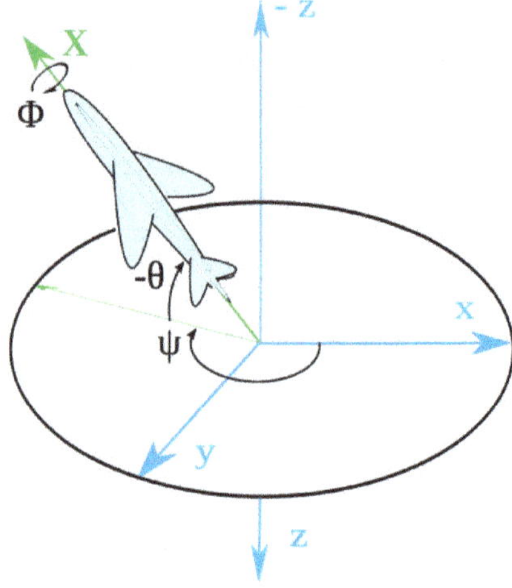

Heading, elevation and bank angles (Z-Y'-X") for an aircraft. The aircraft's pitch and yaw axes Y and Z are not shown, and its fixed reference frame xyz has been shifted backwards from its center of gravity (preserving angles) for clarity. Axes named according to the air norm DIN 9300

In ballistics and flight dynamics, axes conventions are standardized ways of establishing the location and orientation of coordinate axes for use as a frame of reference. Mobile objects are normally tracked from an external frame considered fixed. Other frames can be defined on those mobile objects to deal with relative positions for other objects. Finally, attitudes or orientations can be described by a relationship between the external frame and the one defined over the mobile object.

The orientation of a vehicle is normally referred to as *attitude*. It is described normally by the orientation of a frame fixed in the body relative to a fixed reference frame. The attitude is described by *attitude coordinates*, and consists of at least three coordinates.

While from a geometrical point of view the different methods to describe orientations are defined using only some reference frames, in engineering applications it is important also to describe how these frames are attached to the lab and the body in motion.

Due to the special importance of international conventions in air vehicles, several organizations have published standards to be followed. For example, German DIN has published the DIN 9300 norm for aircraft (adopted by ISO as ISO 1151–2:1985).

Earth Bounded Axes Conventions

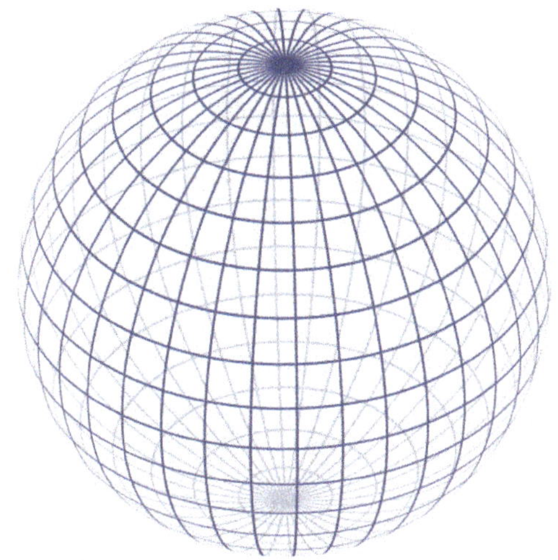

Representation of the earth with parallels and meridians

Ground Reference Frames: ENU and NED

Basically, as lab frame or reference frame, there are two kinds of conventions for the frames (sometimes named LVLH, local vertical, local horizontal):

- East, North, Up, referred as ENU
- North, East, Down, referred as NED, used specially in aerospace

These frames are location dependent. For movements around the globe, like air or sea navigation, the frames are defined as tangent to the lines of coordinates.

- East-West tangent to parallels,
- North-South tangent to meridians, and
- Up-Down in the direction to the center of the earth (when using a spherical Earth simplification), or in the direction normal to the local tangent plane (using an oblate spheroidal or geodetic ellipsoidal model of the earth) which does

not generally pass through the center of the Earth.

To establish a standard convention to describe attitudes, it is required to establish at least the axes of the reference system and the axes of the rigid body or vehicle. When an ambiguous notation system is used (such as Euler angles) also the used convention should be stated. Nevertheless, most used notations (matrices and quaternions) are unambiguous.

Tait–Bryan angles are often used to describe a vehicle's attitude with respect to a chosen reference frame, though any other notation can be used. The positive x-axis in vehicles points always in the direction of movement. For positive y- and z-axis, we have to face two different conventions:

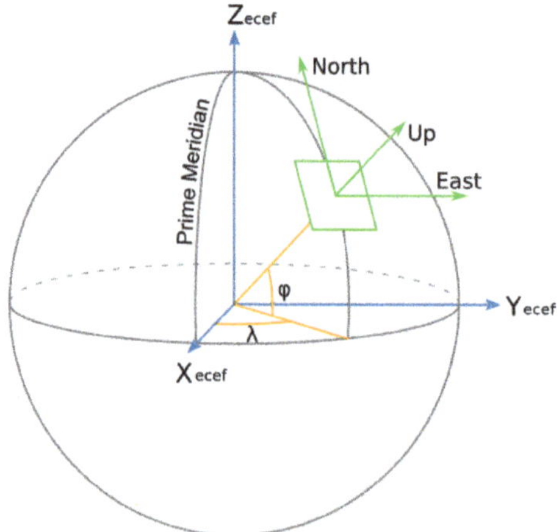

Earth Centred Earth Fixed and East, North, Up coordinates.

- In case of land vehicles like cars, tanks etc., which use the ENU-system (East-North-Up) as external reference (*world frame*), the vehicle's positive y- or pitch axis always points to its left, and the positive z- or yaw axis always points up.
- By contrast, in case of air and sea vehicles like submarines, ships, airplanes etc., which use the NED-system (North-East-Down) as external reference (*world frame*), the vehicle's positive y- or pitch axis always points to its right, and its positive z- or yaw axis always points down.
- Finally, in case of space vehicles like space shuttles etc., a modification of the latter convention is used, where the vehicle's positive y- or pitch axis again always points to its right, and its positive z- or yaw axis always points down, but "down" now may have two different meanings: If a so-called *local frame* is used as external reference, its positive z-axis points "down" to the center of the earth as it does in case of the earlier mentioned NED-system, but if the *inertial frame* is used as reference, its positive z-axis will point now to the North Celestial Pole, and its positive x-axis to the Vernal Equinox or some other reference meridian.

Frames Mounted on Vehicles

Specially for aircraft, these frames do not need to agree with the earth-bound frames in the up-down line. It must be agreed what ENU and NED mean in this context.

Conventions for Land Vehicles

For land vehicles it is rare to describe their complete orientation, except when speaking about electronic stability control or satellite navigation. In this case, the convention is normally the one of the adjacent drawing, where RPY stands for roll-pitch-yaw.

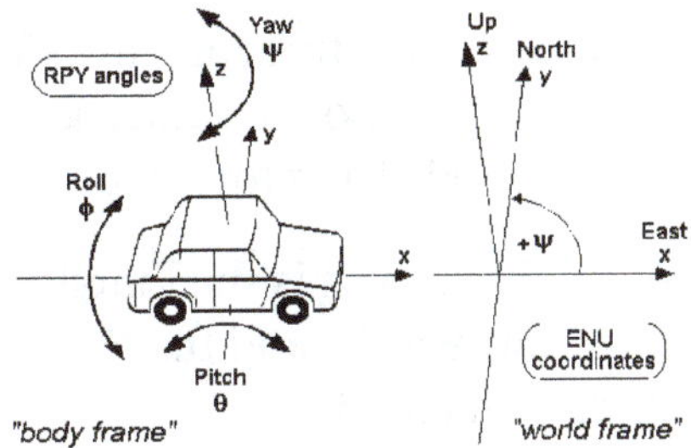

RPY angles of cars and other land vehicles

Conventions for Sea Vehicles

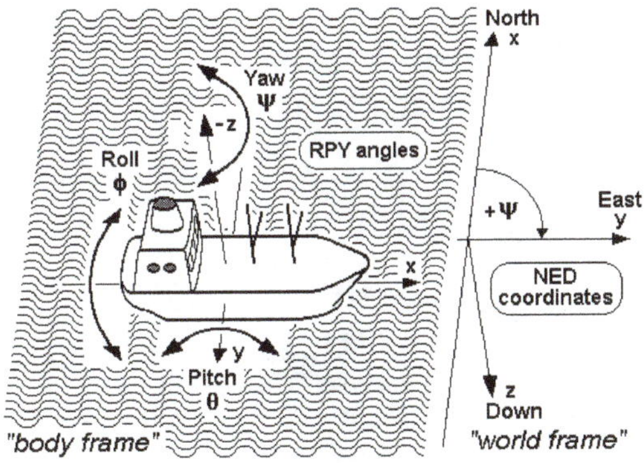

RPY angles of ships and other sea vehicles

As well as aircraft, the same terminology is used for the motion of ships and boats. It is interesting to note that some words commonly used were introduced in maritime navigation. For example, the *yaw* angle or heading, has a nautical origin, with the meaning of “bending out of the course”. Etymologically, it is related with the verb ‘to go’. It is related to the concept of bearing. It is typically assigned the shorthand notation ψ.

Conventions for Aircraft Local Reference Frames

Coordinates to describe an aircraft attitude (Heading, Elevation and Bank) are normally given relative to a reference control frame located in a control tower, and therefore ENU, relative to the position of the control tower on the earth surface.

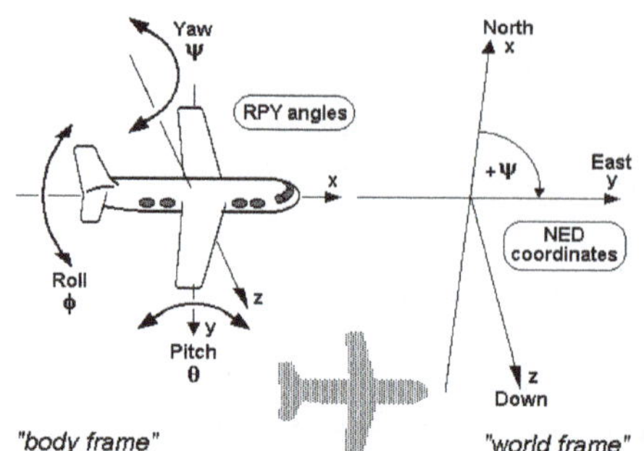

RPY angles of airplanes and other air vehicles

Coordinates to describe observations made from an aircraft are normally given relative to its intrinsic axes, but normally using as positive the coordinate pointing downwards, where the interesting points are located. Therefore, they are normally NED.

These axes are normally taken so that X axis is the longitudinal axis pointing ahead, Z axis is the vertical axis pointing downwards, and the Y axis is the lateral one, pointing in such a way that the frame is right handed.

The *motion* of an aircraft is often described in terms of rotation about these axes, so rotation about the *X*-axis is called rolling, rotation about the *Y*-axis is called pitching, and rotation about the *Z*-axis is called yawing.

Frames for Space Navigation

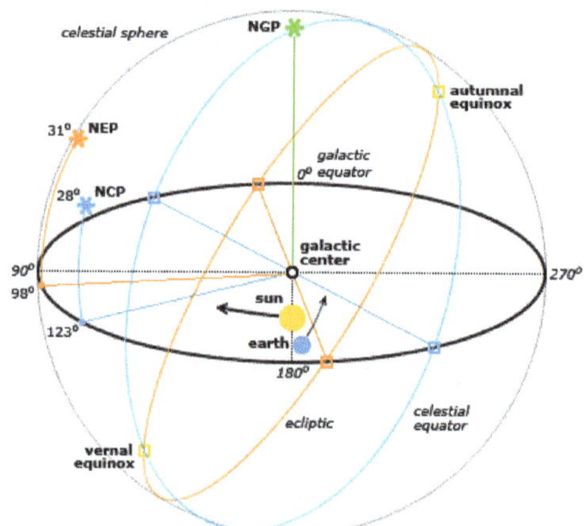

Different reference systems for coordinates in space

For satellites orbiting the earth it is normal to use the Equatorial coordinate system. The projection of the Earth's equator onto the celestial sphere is called the celestial equator. Similarly, the projections of the Earth's north and south geographic poles become the north and south celestial poles, respectively.

Deep space satellites use other Celestial coordinate system, like the Ecliptic coordinate system.

Local Conventions for Space Ships as Satellites

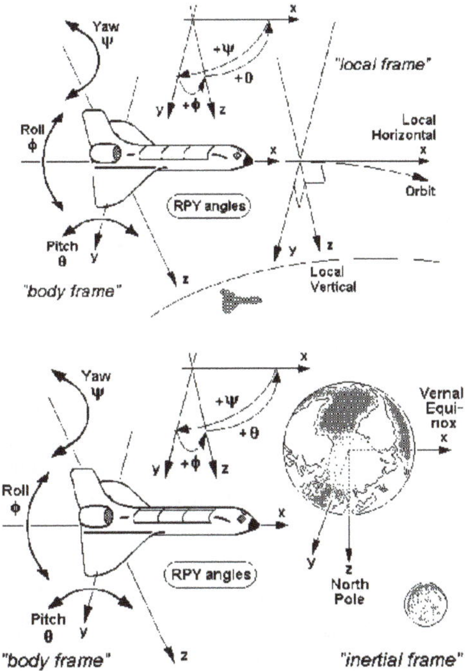

RPY angles of space shuttles and other space vehicles, first using a local frame as reference and second using an inertial frame as reference.

If the goal is to keep the shuttle during its orbits in a constant attitude with respect to the sky, e.g. in order to perform certain astronomical observations, the preferred reference is the *inertial frame*, and the RPY angle vector (0|0|0) describes an attitude then, where the shuttle's wings are kept permanently parallel to the earth's equator, its nose points permanently to the vernal equinox, and its belly towards the northern polar star. (Note that rockets and missiles more commonly follow the conventions for aircraft where the RPY angle vector (0|0|0) points north, rather than toward the vernal equinox).

On the other hand, if the goal is to keep the shuttle during its orbits in a constant attitude with respect to the surface of the earth, the preferred reference will be the *local frame*, with the RPY angle vector (0|0|0) describing an attitude where the shuttle's wings are parallel to the earth's surface, its nose points to its heading, and its belly down towards the centre of the earth.

Frames used to Describe Attitudes

Normally the frames used to describe a vehicle's local observations are the same frames used to describe its attitude respect the ground tracking stations. i.e. if an ENU frame is used in a tracking station, also ENU frames are used onboard and these frames are also used to refer local observations.

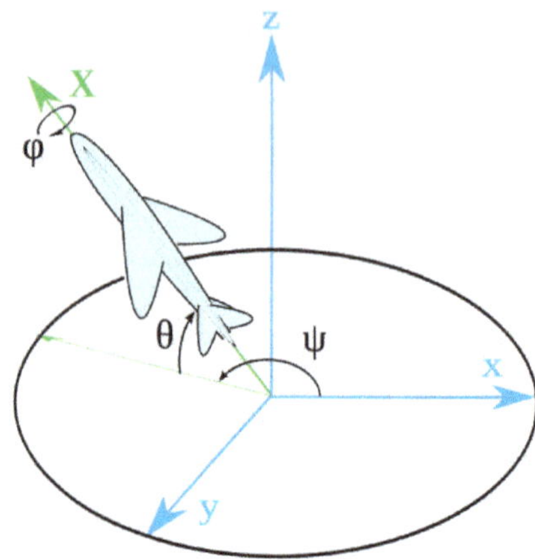

Heading, elevation and bank angles (Z-Y'-X") for an aircraft using onboard ENU axes both onboard and for the ground tracking station. The fixed reference frame x-y-z represents such a tracking station. Onboard axes Y and Z are not shown. X shown in green color.

An important case in which this does not apply is aircraft. Aircraft observations are performed downwards and therefore normally NED axes convention applies. Nevertheless, when attitudes respect ground stations are given, a relationship between the local earth-bound frame and the onboard ENU frame is used.

Map Projection

A medieval depiction of the Ecumene (1482, Johannes Schnitzer, engraver), constructed after the coordinates in Ptolemy's *Geography* and using his second map projection

Commonly, a map projection is a systematic transformation of the latitudes and longitudes of locations on the surface of a sphere or an ellipsoid into locations on a plane. Map projections are necessary for creating maps. All map projections distort the surface in some fashion. Depending on the purpose of the map, some distortions are acceptable and others are not; therefore, different map projections exist in order to preserve some properties of the sphere-like body at the expense of other properties. There is no limit to the number of possible map projections.

More generally, the surfaces of planetary bodies can be mapped even if they are too irregular to be modeled well with a sphere or ellipsoid. Even more generally, projections are the subject of several pure mathematical fields, including differential geometry and projective geometry. However, "map projection" refers specifically to a cartographic projection.

Background

Maps can be more useful than globes in many situations: they are more compact and easier to store; they readily accommodate an enormous range of scales; they are viewed easily on computer displays; they can facilitate measuring properties of the terrain being mapped; they can show larger portions of the Earth's surface at once; and they are cheaper to produce and transport. These useful traits of maps motivate the development of map projections.

However, Carl Friedrich Gauss's Theorema Egregium proved that a sphere's surface cannot be represented on a plane without distortion. The same applies to other reference surfaces used as models for the Earth. Since any map projection is a representation of one of those surfaces on a plane, all map projections distort. Every distinct map projection distorts in a distinct way. The study of map projections is the characterization of these distortions.

Projection is not limited to perspective projections, such as those resulting from casting a shadow on a screen, or the rectilinear image produced by a pinhole camera on a flat film plate. Rather, any mathematical function transforming coordinates from the curved surface to the plane is a projection. Few projections in actual use are perspective.

For simplicity, most of this article assumes that the surface to be mapped is that of a sphere. In reality, the Earth and other large celestial bodies are generally better modeled as oblate spheroids, whereas small objects such as asteroids often have irregular shapes. These other surfaces can be mapped as well. Therefore, more generally, a map projection is any method of "flattening" into a plane a continuous curved surface.

Metric Properties of Maps

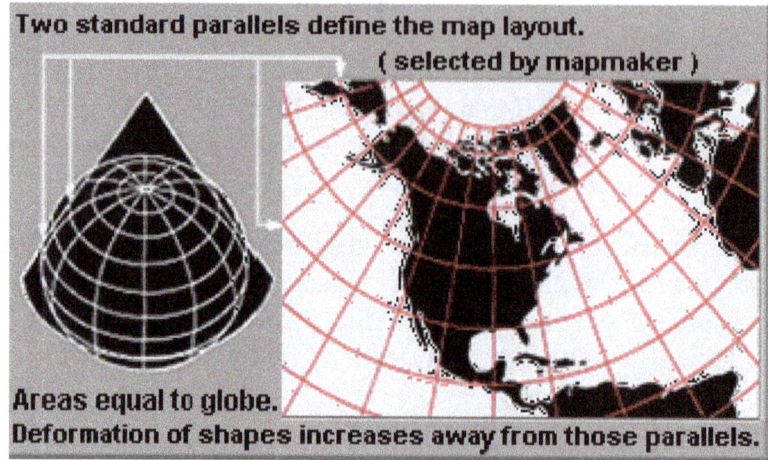

An Albers projection shows areas accurately, but distorts shapes.

Many properties can be measured on the Earth's surface independent of its geography. Some of these properties are:

- Area
- Shape
- Direction
- Bearing
- Distance
- Scale

Map projections can be constructed to preserve at least one of these properties, though only in a limited way for most. Each projection preserves or compromises or approximates basic metric properties in different ways. The purpose of the map determines which projection should form the base for the map. Because many purposes exist for maps, many projections have been created to suit those purposes.

Another consideration in the configuration of a projection is its compatibility with data sets to be used on the map. Data sets are geographic information; their collection depends on the chosen datum (model) of the Earth. Different datums assign slightly different coordinates to the same location, so in large scale maps, such as those from national mapping systems, it is important to match the datum to the projection. The slight differences in coordinate assignation between different datums is not a concern for world maps or other vast territories, where such differences get shrunk to imperceptibility.

Which Projection is Best?

The mathematics of projection do not permit any particular map projection to be "best" for everything. Something will always get distorted. Therefore, a diversity of projections exists to service the many uses of maps and their vast range of scales.

Modern national mapping systems typically employ a transverse Mercator or close variant for large-scale maps in order to preserve conformality and low variation in scale over small areas. For smaller-scale maps, such as those spanning continents or the entire world, many projections are in common use according to their fitness for the purpose.

Thematic maps normally require an equal area projection so that phenomena per unit area are shown in correct proportion. However, representing area ratios correctly necessarily distorts shapes more than many maps that are not equal-area. Hence reference maps of the world often appear on compromise projections instead. Due to distortions inherent in any map of the world, the choice of projection becomes largely one of aesthetics.

The Mercator projection, developed for navigational purposes, has often been used in world maps where other projections would have been more appropriate. This problem has long been recognized even outside professional circles. For example, a 1943 *New York Times* editorial states:

The time has come to discard [the Mercator] for something that represents the continents and directions less deceptively... Although its usage... has diminished... it is still highly popular as a wall map apparently in part because, as a rectangular map, it fills a rectangular wall space with more map, and clearly because its familiarity breeds more popularity.

A controversy in the 1980s over the Peters map motivated the American Cartographic Association (now Cartography and Geographic Information Society) to produce a series of booklets (including *Which Map Is Best*) designed to educate the public about map projections and distortion in maps. In 1989 and 1990, after some internal debate, seven North American geographic organizations adopted a resolution recommending against using any rectangular projection (including Mercator and Gall–Peters) for reference maps of the world.

Distortion

The classical way of showing the distortion inherent in a projection is to use Tissot's indicatrix. For a given point, using the scale factor h along the meridian, the scale factor k along the parallel, and the angle θ' between them, Nicolas Tissot described how to construct an ellipse that characterizes the amount and orientation of the components of distortion. By spacing the ellipses regularly along the meridians and parallels, the network of indicatrices shows how distortion varies across the map.

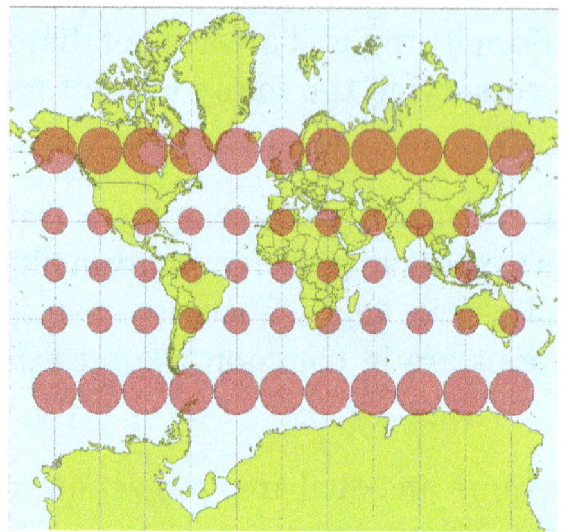

Tissot's Indicatrices on the Mercator projection

Construction of a Map Projection

The creation of a map projection involves two steps:

1. Selection of a model for the shape of the Earth or planetary body (usually choosing between a sphere or ellipsoid). Because the Earth's actual shape is irregular, information is lost in this step.

2. Transformation of geographic coordinates (longitude and latitude) to Cartesian (x,y) or polar plane coordinates. Cartesian coordinates normally have a simple relation to eastings and northings defined on a grid superimposed on the projection.

Some of the simplest map projections are literal projections, as obtained by placing a light source at some definite point relative to the globe and projecting its features onto a specified surface. This is not the case for most projections, which are defined only in terms of mathematical formulae that have no direct geometric interpretation.

Choosing a Projection Surface

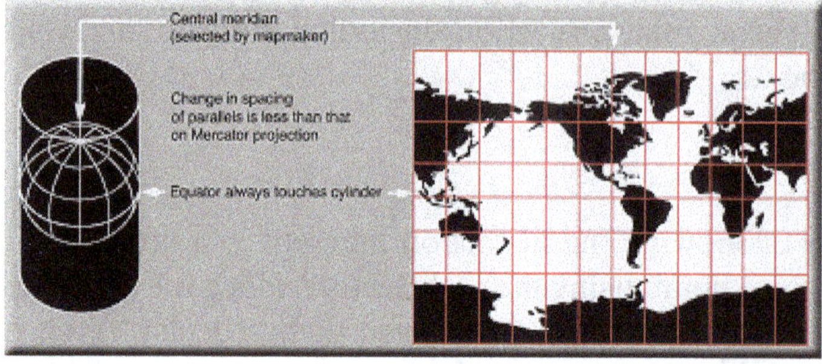

A Miller cylindrical projection maps the globe onto a cylinder.

A surface that can be unfolded or unrolled into a plane or sheet without stretching, tearing or shrinking is called a *developable surface*. The cylinder, cone and the plane are all developable surfaces. The sphere and ellipsoid do not have developable surfaces, so any projection of them onto a plane will have to distort the image. (To compare, one cannot flatten an orange peel without tearing and warping it.)

One way of describing a projection is first to project from the Earth's surface to a developable surface such as a cylinder or cone, and then to unroll the surface into a plane. While the first step inevitably distorts some properties of the globe, the developable surface can then be unfolded without further distortion.

Aspect of the Projection

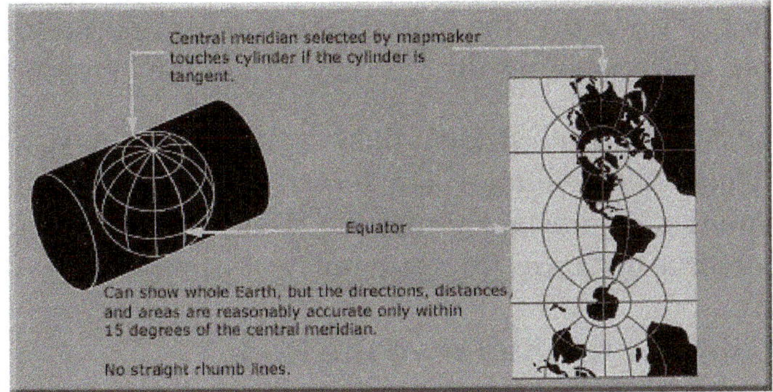

This transverse Mercator projection is mathematically the same as a standard Mercator, but oriented around a different axis.

Once a choice is made between projecting onto a cylinder, cone, or plane, the aspect of the shape must be specified. The aspect describes how the developable surface is placed relative to the globe: it may be *normal* (such that the surface's axis of symmetry coincides with the Earth's axis), *transverse* (at right angles to the Earth's axis) or *oblique* (any angle in between).

Notable Lines

The developable surface may also be either *tangent* or *secant* to the sphere or ellipsoid. Tangent means the surface touches but does not slice through the globe; secant means the surface does slice through the globe. Moving the developable surface away from contact with the globe never preserves or optimizes metric properties, so that possibility is not discussed further here.

Tangent and secant lines (*standard lines*) are represented undistorted. If these lines are a parallel of latitude, as in conical projections, it is called a *standard parallel*. The *central meridian* is the meridian to which the globe is rotated before projecting. The central meridian (usually written λ_0) and a parallel of origin (usually written φ_0) are often used to define the origin of the map projection.

Scale

A globe is the only way to represent the earth with constant scale throughout the entire map in all directions. A map cannot achieve that property for any area, no matter how small. It can, however, achieve constant scale along specific lines.

Some possible properties are:

- The scale depends on location, but not on direction. This is equivalent to preservation of angles, the defining characteristic of a conformal map.
- Scale is constant along any parallel in the direction of the parallel. This applies for any cylindrical or pseudocylindrical projection in normal aspect.
- Combination of the above: the scale depends on latitude only, not on longitude or direction. This applies for the Mercator projection in normal aspect.
- Scale is constant along all straight lines radiating from a particular geographic location. This is the defining characteristic of an equidistant projection such as the Azimuthal equidistant projection. There are also projections (Maurer, Close) where true distances from *two* points are preserved.

Choosing a Model for the Shape of the Body

Projection construction is also affected by how the shape of the Earth or planetary body is approximated. In the following section on projection categories, the earth is taken as a sphere in order to simplify the discussion. However, the Earth's actual shape is closer to an oblate ellipsoid. Whether spherical or ellipsoidal, the principles discussed hold without loss of generality.

Selecting a model for a shape of the Earth involves choosing between the advantages and disadvantages of a sphere versus an ellipsoid. Spherical models are useful for small-scale maps such as world atlases and globes, since the error at that scale is not usually noticeable or important enough to justify using the more complicated ellipsoid. The ellipsoidal model is commonly used to construct topographic maps and for other large- and medium-scale maps that need to accurately depict the land surface. Auxiliary latitudes are often employed in projecting the ellipsoid.

A third model is the geoid, a more complex and accurate representation of Earth's shape coincident with what mean sea level would be if there were no winds, tides, or land. Compared to the best fitting ellipsoid, a geoidal model would change the characterization of important properties such as distance, conformality and equivalence. Therefore, in geoidal projections that preserve such properties, the mapped graticule would deviate from a mapped ellipsoid's graticule. Normally the geoid is not used as an Earth model for projections, however, because Earth's shape is very regular, with the undulation of the geoid amounting to less than 100 m from the ellipsoidal model out of the 6.3 million m Earth radius. For irregular planetary bodies such as asteroids, however, sometimes models analogous to the geoid are used to project maps from.

Classification

A fundamental projection classification is based on the type of projection surface onto which the globe is conceptually projected. The projections are described in terms of placing a gigantic surface in contact with the earth, followed by an implied scaling operation. These surfaces are cylindrical (e.g. Mercator), conic (e.g. Albers), or azimuthal or plane (e.g. stereographic). Many mathematical projections, however, do not neatly fit into any of these three conceptual projection methods. Hence other peer categories have been described in the literature, such as pseudoconic, pseudocylindrical, pseudoazimuthal, retroazimuthal, and polyconic.

Another way to classify projections is according to properties of the model they preserve. Some of the more common categories are:

- Preserving direction (*azimuthal or zenithal*), a trait possible only from one or two points to every other point
- Preserving shape locally (*conformal* or *orthomorphic*)
- Preserving area (*equal-area* or *equiareal* or *equivalent* or *authalic*)
- Preserving distance (*equidistant*), a trait possible only between one or two points and every other point
- Preserving shortest route, a trait preserved only by the gnomonic projection

Because the sphere is not a developable surface, it is impossible to construct a map projection that is both equal-area and conformal.

Projections by Surface

The three developable surfaces (plane, cylinder, cone) provide useful models for understanding, describing, and developing map projections. However, these models are limited in two fundamental ways. For one thing, most world projections in actual use do not fall into any of those categories. For another thing, even most projections that do fall into those categories are not naturally attainable through physical projection. As L.P. Lee notes,

No reference has been made in the above definitions to cylinders, cones or planes. The projections are termed cylindric or conic because they can be regarded as developed on a cylinder or a cone, as the case may be, but it is as well to dispense with picturing cylinders and cones, since they have given rise to much misunderstanding. Particularly is this so with regard to the conic projections with two standard parallels: they may be regarded as developed on cones, but they are cones which bear no simple relationship to the sphere. In reality, cylinders and cones provide us with convenient descriptive terms, but little else.

Lee's objection refers to the way the terms *cylindrical, conic,* and *planar* (azimuthal) have been abstracted in the field of map projections. If maps were projected as in light shining through a globe onto a developable surface, then the spacing of parallels would follow a very limited set of possibilities. Such a cylindrical projection (for example) is one which:

1. Is rectangular;
2. Has straight vertical meridians, spaced evenly;
3. Has straight parallels symmetrically placed about the equator;
4. Has parallels constrained to where they fall when light shines through the globe onto the cylinder, with the light source someplace along the line formed by the intersection of the prime meridian with the equator, and the center of the sphere.

(If you rotate the globe before projecting then the parallels and meridians will not necessarily still be straight lines. Rotations are normally ignored for the purpose of classification.)

Where the light source emanates along the line described in this last constraint is what yields the differences between the various "natural" cylindrical projections. But the term *cylindrical* as used in the field of map projections relaxes the last constraint entirely. Instead the parallels can be placed according to any algorithm the designer has decided suits the needs of the map. The famous Mercator projection is one in which the placement of parallels does not arise by "projection"; instead parallels are placed how they need to be in order to satisfy the property that a course of constant bearing is always plotted as a straight line.

Cylindrical

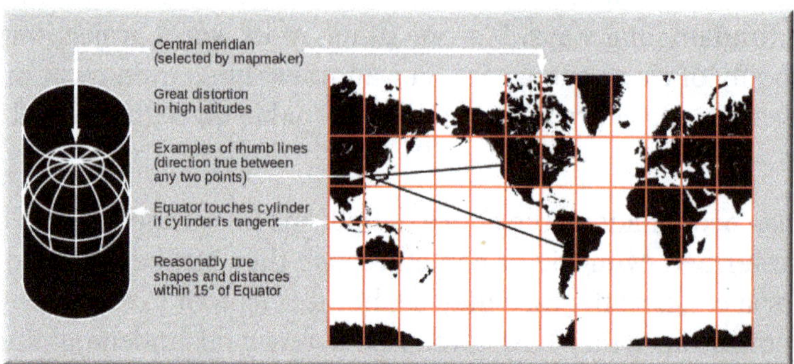

The Mercator projection shows rhumbs as straight lines. A rhumb is a course of constant bearing. Bearing is the compass direction of movement.

The term “normal cylindrical projection” is used to refer to any projection in which meridians are mapped to equally spaced vertical lines and circles of latitude (parallels) are mapped to horizontal lines.

The mapping of meridians to vertical lines can be visualized by imagining a cylinder whose axis coincides with the Earth’s axis of rotation. This cylinder is wrapped around the Earth, projected onto, and then unrolled.

By the geometry of their construction, cylindrical projections stretch distances east-west. The amount of stretch is the same at any chosen latitude on all cylindrical projections, and is given by the secant of the latitude as a multiple of the equator’s scale. The various cylindrical projections are distinguished from each other solely by their north-south stretching (where latitude is given by φ):

- North-south stretching equals east-west stretching (sec φ): The east-west scale matches the north-south scale: conformal cylindrical or Mercator; this distorts areas excessively in high latitudes.
- North-south stretching grows with latitude faster than east-west stretching ($\sec^2 \varphi$): The cylindric perspective (or central cylindrical) projection; unsuitable because distortion is even worse than in the Mercator projection.
- North-south stretching grows with latitude, but less quickly than the east-west stretching: such as the Miller cylindrical projection (sec 4φ/5).
- North-south distances neither stretched nor compressed (1): equirectangular projection or “plate carrée”.
- North-south compression equals the cosine of the latitude (the reciprocal of east-west stretching): equal-area cylindrical. This projection has many named specializations differing only in the scaling constant, such as the Gall–Peters or Gall orthographic (undistorted at the 45° parallels), Behrmann (undistorted at the 30° parallels), and Lambert cylindrical equal-area (undistorted at the equator). Since this projection scales north-south distances by the reciprocal of east-west stretching, it preserves area at the expense of shapes.

In the first case (Mercator), the east-west scale always equals the north-south scale. In the second case (central cylindrical), the north-south scale exceeds the east-west scale everywhere away from the equator. Each remaining case has a pair of secant lines—a pair of identical latitudes of opposite sign (or else the equator) at which the east-west scale matches the north-south-scale.

Normal cylindrical projections map the whole Earth as a finite rectangle, except in the first two cases, where the rectangle stretches infinitely tall while retaining constant width.

Pseudocylindrical

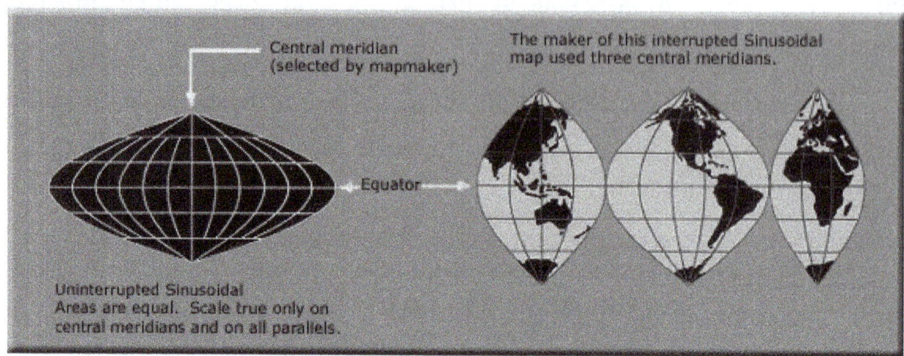

A sinusoidal projection shows relative sizes accurately, but grossly distorts shapes. Distortion can be reduced by "interrupting" the map.

Pseudocylindrical projections represent the *central* meridian as a straight line segment. Other meridians are longer than the central meridian and bow outward away from the central meridian. Pseudocylindrical projections map parallels as straight lines. Along parallels, each point from the surface is mapped at a distance from the central meridian that is proportional to its difference in longitude from the central meridian. On a pseudocylindrical map, any point further from the equator than some other point has a higher latitude than the other point, preserving north-south relationships. This trait is useful when illustrating phenomena that depend on latitude, such as climate. Examples of pseudocylindrical projections include:

- Sinusoidal, which was the first pseudocylindrical projection developed. Vertical scale and horizontal scale are the same throughout, resulting in an equal-area map. On the map, as in reality, the length of each parallel is proportional to the cosine of the latitude. Thus the shape of the map for the whole earth is the region between two symmetric rotated cosine curves. The true distance between two points on the same meridian corresponds to the distance on the map between the two parallels, which is smaller than the distance between the two points on the map. The distance between two points on the same parallel is true. The area of any region is true.
- Collignon projection, which in its most common forms represents each meridian as two straight line segments, one from each pole to the equator.

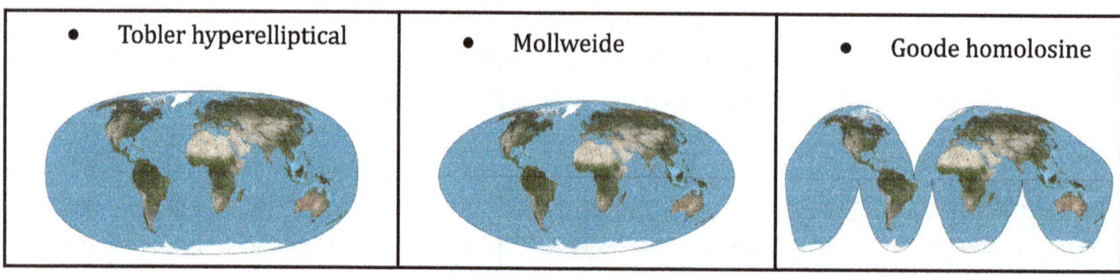

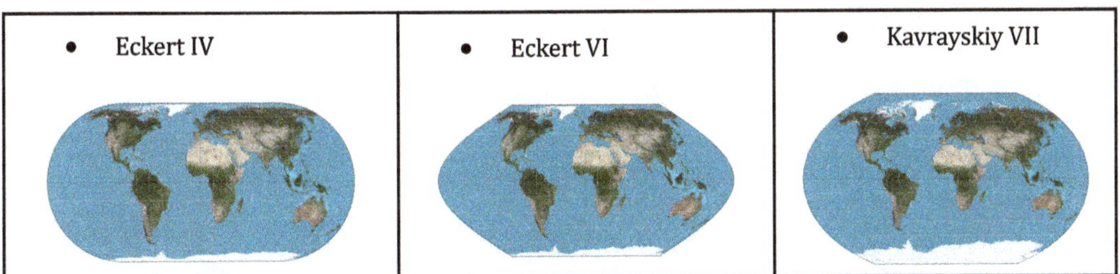

Hybrid

The HEALPix projection combines an equal-area cylindrical projection in equatorial regions with the Collignon projection in polar areas.

Conic

The term "conic projection" is used to refer to any projection in which meridians are mapped to equally spaced lines radiating out from the apex and circles of latitude (parallels) are mapped to circular arcs centered on the apex.

When making a conic map, the map maker arbitrarily picks two standard parallels. Those standard parallels may be visualized as secant lines where the cone intersects the globe—or, if the map maker chooses the same parallel twice, as the tangent line where the cone is tangent to the globe. The resulting conic map has low distortion in scale, shape, and area near those standard parallels. Distances along the parallels to the north of both standard parallels or to the south of both standard parallels are stretched; distances along parallels between the standard parallels are compressed. When a single standard parallel is used, distances along all other parallels are stretched.

The most popular conic maps include:

- Equidistant conic, which keeps parallels evenly spaced along the meridians to preserve a constant distance scale along each meridian, typically the same or similar scale as along the standard parallels.
- Albers conic, which adjusts the north-south distance between non-standard parallels to compensate for the east-west stretching or compression, giving an equal-area map.
- Lambert conformal conic, which adjusts the north-south distance between non-standard parallels to equal the east-west stretching, giving a conformal map.

Pseudoconic

- Bonne
- Werner cordiform, upon which distances are correct from one pole, as well as

along all parallels.

- Continuous American polyconic

Azimuthal (Projections onto a Plane)

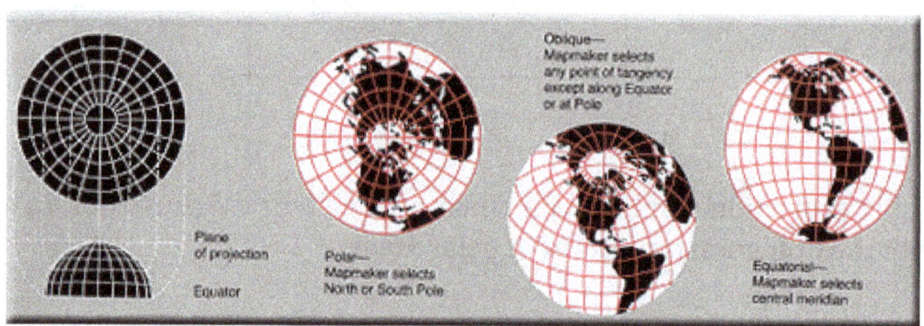

An azimuthal equidistant projection shows distances and directions accurately from the center point, but distorts shapes and sizes elsewhere.

Azimuthal projections have the property that directions from a central point are preserved and therefore great circles through the central point are represented by straight lines on the map. Usually these projections also have radial symmetry in the scales and hence in the distortions: map distances from the central point are computed by a function $r(d)$ of the true distance d, independent of the angle; correspondingly, circles with the central point as center are mapped into circles which have as center the central point on the map.

The mapping of radial lines can be visualized by imagining a plane tangent to the Earth, with the central point as tangent point.

The radial scale is $r'(d)$ and the transverse scale $r(d)/(R \sin d/R)$ where R is the radius of the Earth.

Some azimuthal projections are true perspective projections; that is, they can be constructed mechanically, projecting the surface of the Earth by extending lines from a point of perspective (along an infinite line through the tangent point and the tangent point's antipode) onto the plane:

- The gnomonic projection displays great circles as straight lines. Can be constructed by using a point of perspective at the center of the Earth. $r(d) = c \tan d/R$; so that even just a hemisphere is already infinite in extent.

- The General Perspective projection can be constructed by using a point of perspective outside the earth. Photographs of Earth (such as those from the International Space Station) give this perspective.

- The orthographic projection maps each point on the earth to the closest point on the plane. Can be constructed from a point of perspective an infinite distance from the tangent point; $r(d) = c \sin d/R$. Can display up to a hemisphere on a finite circle. Photographs of Earth from far enough away, such as the Moon, give this perspective.

- The azimuthal conformal projection, also known as the stereographic projection, can be constructed by using the tangent point's antipode as the point of perspective. $r(d) = c \tan d/2R$; the scale is $c/(2R \cos^2 d/2R)$. Can display nearly the entire sphere's surface on a finite circle. The sphere's full surface requires an infinite map.

Other azimuthal projections are not true perspective projections:

- Azimuthal equidistant: $r(d) = cd$; it is used by amateur radio operators to know the direction to point their antennas toward a point and see the distance to it. Distance from the tangent point on the map is proportional to surface distance on the earth (; for the case where the tangent point is the North Pole.
- Lambert azimuthal equal-area. Distance from the tangent point on the map is proportional to straight-line distance through the earth: $r(d) = c \sin d/2R$
- Logarithmic azimuthal is constructed so that each point's distance from the center of the map is the logarithm of its distance from the tangent point on the Earth. $r(d) = c \ln d/d_0$); locations closer than at a distance equal to the constant d_0 are not shown.

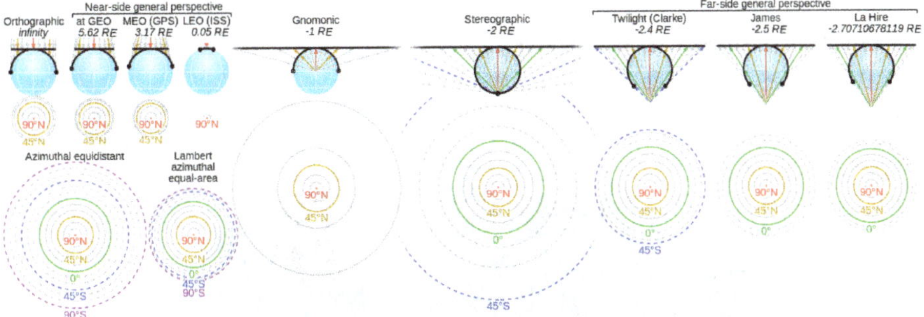

Comparison of some azimuthal projections centred on 90° N at the same scale, ordered by projection altitude in Earth radii.

Projections by Preservation of a Metric Property

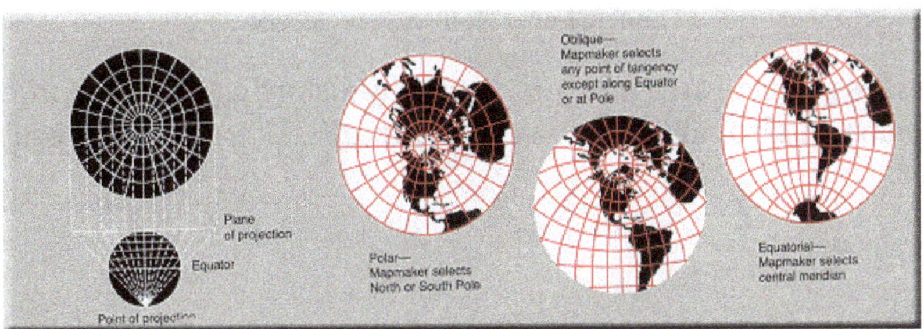

A stereographic projection is conformal and perspective but not equal area or equidistant.

Conformal

Conformal, or orthomorphic, map projections preserve angles locally, implying that they map infinitesimal circles of constant size anywhere on the Earth to infinitesimal circles of varying sizes on the map. In contrast, mappings that are not conformal distort most such small circles into ellipses of distortion. An important consequence of conformality is that relative angles at each point of the map are correct, and the local scale (although varying throughout the map) in every direction around any one point is constant. These are some conformal projections:

- Mercator: Rhumb lines are represented by straight segments
- Transverse Mercator
- Stereographic: Any circle of a sphere, great and small, maps to a circle or straight line.
- Roussilhe
- Lambert conformal conic
- Peirce quincuncial projection
- Adams hemisphere-in-a-square projection
- Guyou hemisphere-in-a-square projection

Equal-area

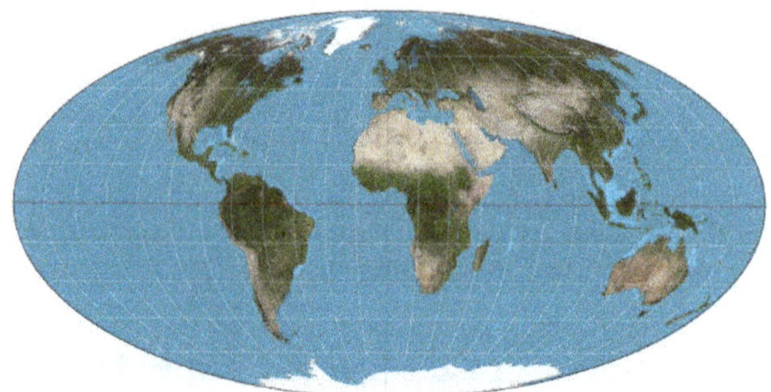

The equal-area Mollweide projection

Equal-area maps preserve area measure, generally distorting shapes in order to do that. Equal-area maps are also called *equivalent* or *authalic*. These are some projections that preserve area:

- Albers conic
- Bonne

- Bottomley
- Collignon
- Cylindrical equal-area
- Eckert II, IV and VI
- Gall orthographic (also known as Gall–Peters, or Peters, projection)
- Goode's homolosine
- Hammer
- Hobo–Dyer
- Lambert azimuthal equal-area
- Lambert cylindrical equal-area
- Mollweide
- Sinusoidal
- Snyder's equal-area polyhedral projection, used for geodesic grids.
- Tobler hyperelliptical
- Werner

Equidistant

A two-point equidistant projection of Eurasia

These are some projections that preserve distance from some standard point or line:

- Equirectangular—distances along meridians are conserved

- Plate carrée—an Equirectangular projection centered at the equator
- Azimuthal equidistant—distances along great circles radiating from centre are conserved
- Equidistant conic
- Sinusoidal—distances along parallels are conserved
- Werner cordiform distances from the North Pole are correct as are the curved distance on parallels
- Soldner
- Two-point equidistant: two "control points" are arbitrarily chosen by the map maker. Distance from any point on the map to each control point is proportional to surface distance on the earth.

Gnomonic

Great circles are displayed as straight lines:

- Gnomonic projection

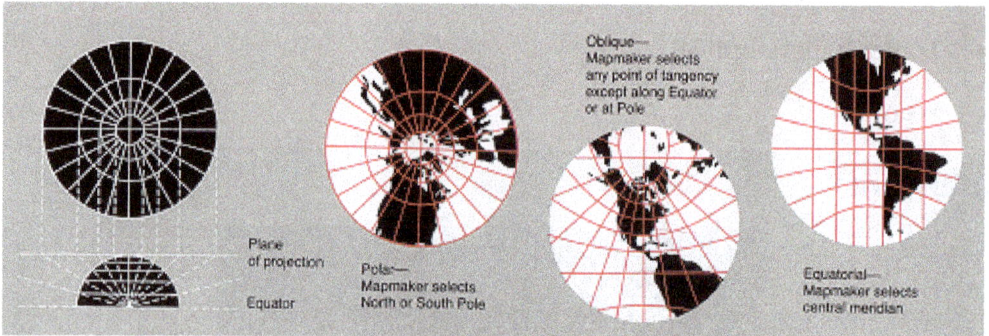

The Gnomonic projection is thought to be the oldest map projection, developed by Thales in the 6th century BC

Retroazimuthal

Direction to a fixed location B (the bearing at the starting location A of the shortest route) corresponds to the direction on the map from A to B:

- Littrow—the only conformal retroazimuthal projection
- Hammer retroazimuthal—also preserves distance from the central point
- Craig retroazimuthal *aka* Mecca or Qibla—also has vertical meridians

Compromise Projections

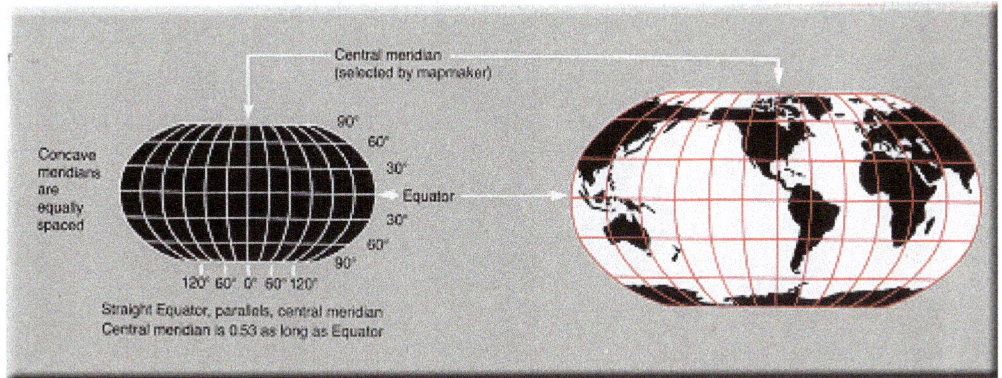

The Robinson projection was adopted by *National Geographic* magazine in 1988 but abandoned by them in about 1997 for the Winkel tripel.

Compromise projections give up the idea of perfectly preserving metric properties, seeking instead to strike a balance between distortions, or to simply make things "look right". Most of these types of projections distort shape in the polar regions more than at the equator. These are some compromise projections:

- Robinson
- van der Grinten
- Miller cylindrical
- Winkel Tripel
- Buckminster Fuller's Dymaxion
- B. J. S. Cahill's Butterfly Map
- Kavrayskiy VII projection
- Wagner VI projection
- Chamberlin trimetric
- Oronce Finé's cordiform

Stereographic Projection

In geometry, the stereographic projection is a particular mapping (function) that projects a sphere onto a plane. The projection is defined on the entire sphere, except at one point: the projection point. Where it is defined, the mapping is smooth and bijective. It is conformal, meaning that it preserves angles. It is neither isometric nor area-preserving: that is, it preserves neither distances nor the areas of figures.

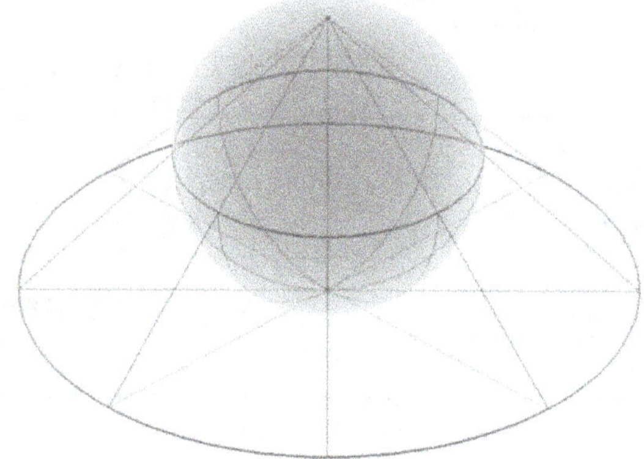

3D illustration of a stereographic projection from the north pole onto a plane below the sphere

Intuitively, then, the stereographic projection is a way of picturing the sphere as the plane, with some inevitable compromises. Because the sphere and the plane appear in many areas of mathematics and its applications, so does the stereographic projection; it finds use in diverse fields including complex analysis, cartography, geology, and photography. In practice, the projection is carried out by computer or by hand using a special kind of graph paper called a stereographic net, shortened to stereonet, or Wulff net.

History

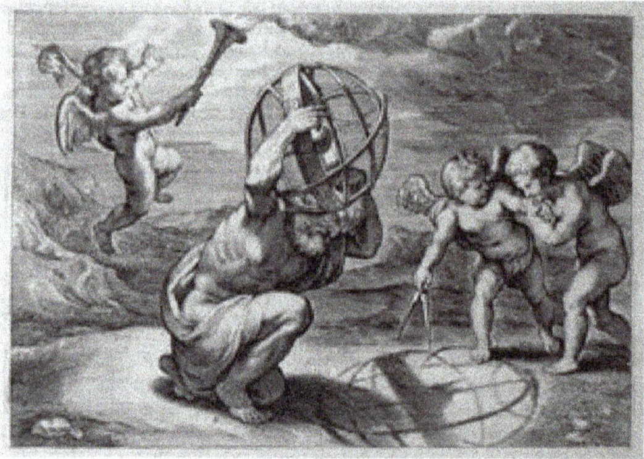

Illustration by Rubens for "Opticorum libri sex philosophis juxta ac mathematicis utiles", by François d'Aguilon. It demonstrates how the projection is computed.

The stereographic projection was known to Hipparchus, Ptolemy and probably earlier to the Egyptians. It was originally known as the planisphere projection. *Planisphaerium* by Ptolemy is the oldest surviving document that describes it. One of its most important uses was the representation of celestial charts. The term *planisphere* is still used to refer to such charts.

In the 16th and 17th century, the equatorial aspect of the stereographic projection was commonly used for maps of the Eastern and Western Hemispheres. It is believed that already the map created in 1507 by Gualterius Lud was in stereograhic projection, as were later the maps of Jean Roze (1542), Rumold Mercator (1595), and many others. In star charts, even this equatorial aspect had been utilised already by the ancient astronomers like Ptolemy.

François d'Aguilon gave the stereographic projection its current name in his 1613 work *Opticorum libri sex philosophis juxta ac mathematicis utiles* (Six Books of Optics, useful for philosophers and mathematicians alike).

In 1695, Edmond Halley, motivated by his interest in star charts, published the first mathematical proof that this map is conformal. He used the recently established tools of calculus, invented by his friend Isaac Newton.

Definition

This section focuses on the projection of the unit sphere from the north pole onto the plane through the equator. Other formulations are treated in later sections.

The unit sphere in three-dimensional space R^3 is the set of points (x, y, z) such that $x^2 + y^2 + z^2 = 1$. Let $N = (0, 0, 1)$ be the "north pole", and let M be the rest of the sphere. The plane $z = 0$ runs through the center of the sphere; the "equator" is the intersection of the sphere with this plane.

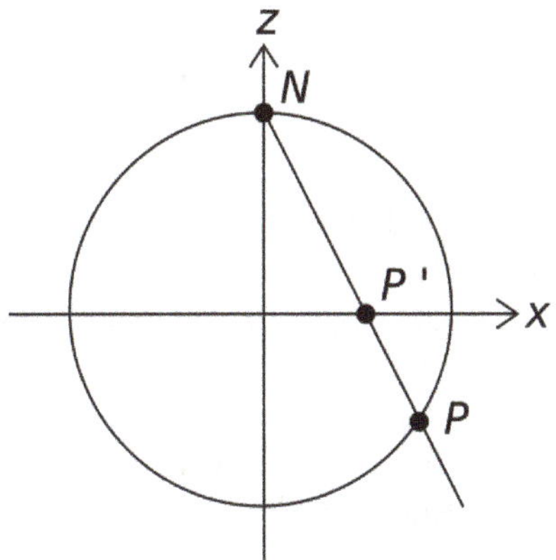

Stereographic projection of the unit sphere from the north pole onto the plane $z = 0$, shown here in cross section

For any point P on M, there is a unique line through N and P, and this line intersects the plane $z = 0$ in exactly one point P'. Define the stereographic projection of P to be this point P' in the plane.

In Cartesian coordinates (x, y, z) on the sphere and (X, Y) on the plane, the projection and its inverse are given by the formulas

$$(X, Y) = \left(\frac{x}{1-z}, \frac{y}{1-z}\right),$$
$$(x, y, z) = \left(\frac{2X}{1+X^2+Y^2}, \frac{2Y}{1+X^2+Y^2}, \frac{-1+X^2+Y^2}{1+X^2+Y^2}\right).$$

In spherical coordinates (φ, θ) on the sphere (with φ the zenith angle, $0 \le \varphi \le \pi$, and θ the azimuth, $0 \le \theta \le 2\pi$) and polar coordinates (R, Θ) on the plane, the projection and its inverse are

$$(R, \Theta) = \left(\frac{\sin\varphi}{1-\cos\varphi}, \theta\right) = \left(\cot\frac{\varphi}{2}, \theta\right),$$

$$(\varphi, \theta) = \left(2\arctan\frac{1}{R}, \Theta\right).$$

Here, φ is understood to have value π when $R = 0$. Also, there are many ways to rewrite these formulas using trigonometric identities. In cylindrical coordinates (r, θ, z) on the sphere and polar coordinates (R, Θ) on the plane, the projection and its inverse are

$$(R, \Theta) = \left(\frac{r}{1-z}, \theta\right),$$
$$(r, \theta, z) = \left(\frac{2R}{1+R^2}, \Theta, \frac{R^2-1}{R^2+1}\right).$$

Properties

The stereographic projection defined in the preceding section sends the "south pole" (0, 0, −1) of the unit sphere to (0, 0), the equator to the unit circle, the southern hemisphere to the region inside the circle, and the northern hemisphere to the region outside the circle.

The projection is not defined at the projection point $N = (0, 0, 1)$. Small neighborhoods of this point are sent to subsets of the plane far away from (0, 0). The closer P is to (0, 0, 1), the more distant its image is from (0, 0) in the plane. For this reason it is common to speak of (0, 0, 1) as mapping to "infinity" in the plane, and of the sphere as completing the plane by adding a "point at infinity". This notion finds utility in projective geometry and complex analysis. On a merely topological level, it illustrates how the sphere is homeomorphic to the one-point compactification of the plane.

In Cartesian coordinates a point $P(x, y, z)$ on the sphere and its image $P'(X, Y)$ on the plane either both are rational points or none of them:

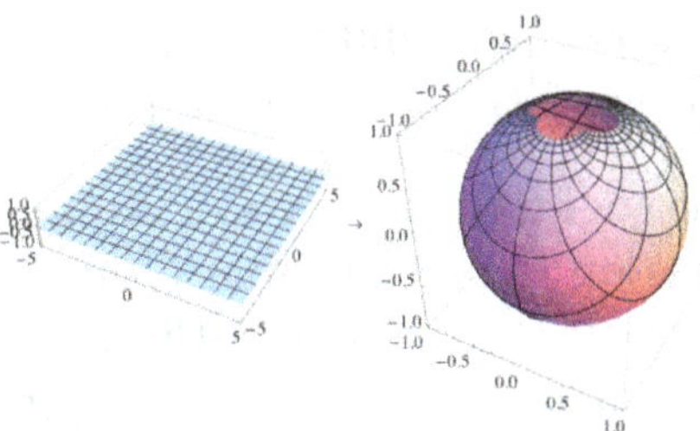

A Cartesian grid on the plane appears distorted on the sphere. The grid lines are still perpendicular, but the areas of the grid squares shrink as they approach the north pole.

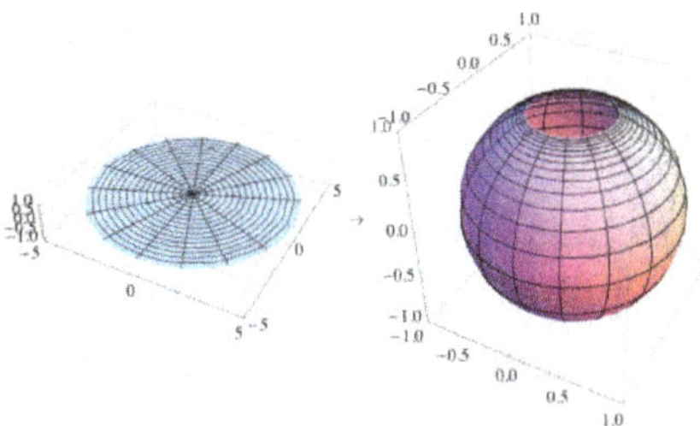

A polar grid on the plane appears distorted on the sphere. The grid curves are still perpendicular, but the areas of the grid sectors shrink as they approach the north pole.

$$\mathrm{P} \in \mathbb{Q}^3 \Leftrightarrow \mathrm{P}' \in \mathbb{Q}^2$$

Stereographic projection is conformal, meaning that it preserves the angles at which curves cross each other. On the other hand, stereographic projection does not preserve area; in general, the area of a region of the sphere does not equal the area of its projection onto the plane. The area element is given in (X, Y) coordinates by

$$dA = \frac{4}{(1+X^2+Y^2)^2}\,dX\,dY.$$

Along the unit circle, where $X^2 + Y^2 = 1$, there is no infinitesimal distortion of area. Near (0, 0) areas are distorted by a factor of 4, and near infinity areas are distorted by arbitrarily small factors.

The metric is given in (X, Y) coordinates by

$$\frac{}{(1+X^2+Y^2)^2}\,dX\,dY$$

and is the unique formula found in Bernhard Riemann's *Habilitationsschrift* on the foundations of geometry, delivered at Göttingen in 1854, and entitled *Über die Hypothesen welche der Geometrie zu Grunde liegen.*

No map from the sphere to the plane can be both conformal and area-preserving. If it were, then it would be a local isometry and would preserve Gaussian curvature. The sphere and the plane have different Gaussian curvatures, so this is impossible.

The conformality of the stereographic projection implies a number of convenient geometric properties. Circles on the sphere that do *not* pass through the point of projection are projected to circles on the plane. Circles on the sphere that *do* pass through the point of projection are projected to straight lines on the plane. These lines are sometimes thought of as circles through the point at infinity, or circles of infinite radius.

All lines in the plane, when transformed to circles on the sphere by the inverse of stereographic projection, meet at the projection point. Parallel lines, which do not intersect in the plane, are transformed to circles tangent at projection point. Intersecting lines are transformed to circles that intersect transversally at two points in the sphere, one of which is the projection point. (Similar remarks hold about the real projective plane, but the intersection relationships are different there.)

The sphere, with various loxodromes shown in distinct colors

The loxodromes of the sphere map to curves on the plane of the form

$$R = e^{\frac{\Theta}{a}},$$

where the parameter a measures the "tightness" of the loxodrome. Thus loxodromes correspond to logarithmic spirals. These spirals intersect radial lines in the plane at equal angles, just as the loxodromes intersect meridians on the sphere at equal angles.

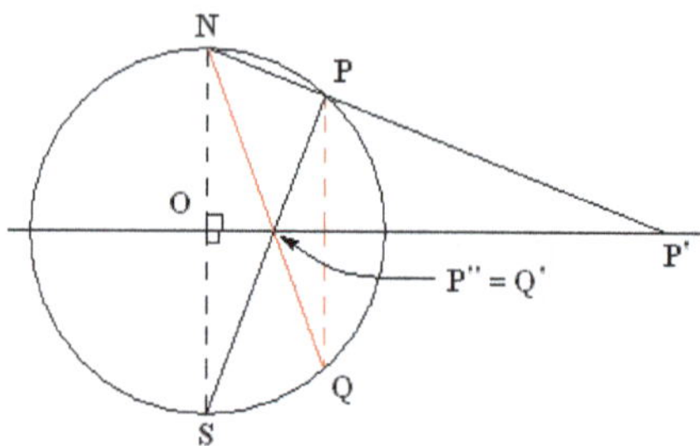

The stereographic projection relates to the plane inversion in a simple way. Let P and

Q be two points on the sphere with projections P′ and Q′ on the plane. Then P′ and Q′ are inversive images of each other in the image of the equatorial circle if and only if P and Q are reflections of each other in the equatorial plane.

In other words, if:

- *P* is a point on the sphere, but not a 'north pole' *N* and not its antipode, the 'south pole' *S*,
- *P′* is the image of *P* in a stereographic projection with the projection point *N* and
- *P″* is the image of *P* in a stereographic projection with the projection point *S*,

then *P′* and *P″* are inversive images of each other in the unit circle.

Wulff Net

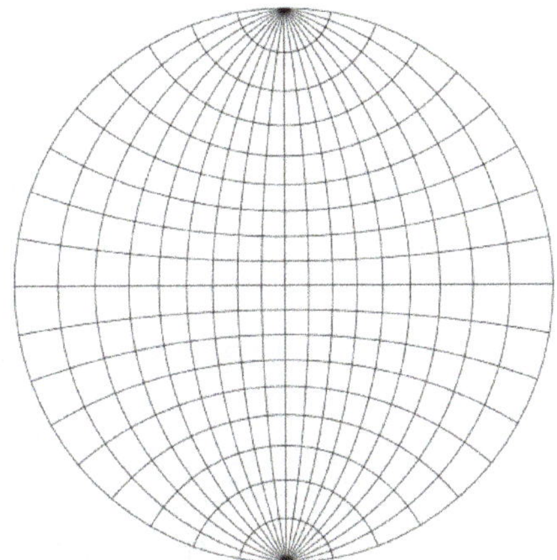

Wulff net or stereonet, used for making plots of the stereographic projection by hand

Stereographic projection plots can be carried out by a computer using the explicit formulas given above. However, for graphing by hand these formulas are unwieldy. Instead, it is common to use graph paper designed specifically for the task. This special graph paper is called a stereonet or Wulff net, after the Russian mineralogist George (Yuri Viktorovich) Wulff.

To make a Wulff net, one places a grid of parallels and meridians on the hemisphere, and then stereographically projects these curves to the disk. Depending on the particular projection used, the parallels and meridians may or may not match those usually encountered in geography. For example, the figure at left is constructed using the conventions of the Definition section above. Because the projection point is (0, 0, 1), the Wulff net depicts the southern hemisphere $z \leq 0$. The equator plots at the circular boundary of the Wulff net, and the south pole plots at the center of the Wulff net. The parallels are chosen to be small circles about the y-axis, and all of the meridians pass through (0, 1, 0) and (0, −1, 0).

In the figure, the area-distorting property of the stereographic projection can be seen by comparing a grid sector near the center of the net with one at the far right of the net. The two sectors have equal areas on the sphere. On the disk, the latter has nearly four times the area of the former. If one uses finer and finer grids on the sphere, then the ratio of the areas approaches exactly 4.

On the Wulff net, the images of the parallels and meridians intersect at right angles. This orthogonality property is a consequence of the angle-preserving property of the stereoscopic projection. (However, the angle-preserving property is stronger than this property. Not all projections that preserve the orthogonality of parallels and meridians are angle-preserving.)

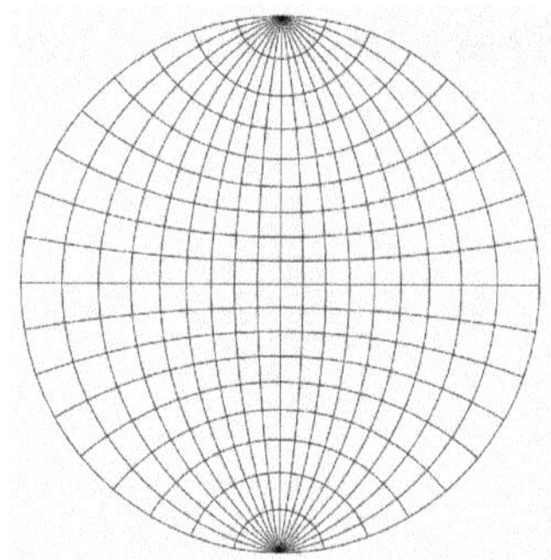

Illustration of steps 1–4 for plotting a point on a Wulff net

For an example of the use of the Wulff net, imagine two copies of it on thin paper, one atop the other, aligned and tacked at their mutual center. Let P be the point on the lower unit hemisphere whose spherical coordinates are (140°, 60°) and whose Cartesian coordinates are (0.321, 0.557, −0.766). This point lies on a line oriented 60° counterclockwise from the positive x-axis (or 30° clockwise from the positive y-axis) and 50° below the horizontal plane $z = 0$. Once these angles are known, there are four steps to plotting P:

1. Using the grid lines, which are spaced 10° apart in the figures here, mark the point on the edge of the net that is 60° counterclockwise from the point (1, 0) (or 30° clockwise from the point (0, 1)).
2. Rotate the top net until this point is aligned with (1, 0) on the bottom net.
3. Using the grid lines on the bottom net, mark the point that is 50° toward the center from that point.
4. Rotate the top net oppositely to how it was oriented before, to bring it back into alignment with the bottom net. The point marked in step 3 is then the projection that we wanted.

To plot other points, whose angles are not such round numbers as 60° and 50°, one must visually interpolate between the nearest grid lines. It is helpful to have a net with finer spacing than 10°. Spacings of 2° are common.

To find the central angle between two points on the sphere based on their stereographic plot, overlay the plot on a Wulff net and rotate the plot about the center until the two points lie on or near a meridian. Then measure the angle between them by counting grid lines along that meridian.

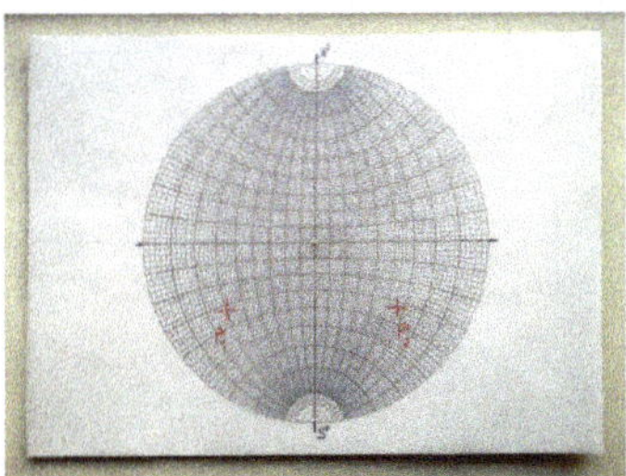

Two points P_1 and P_2 are drawn on a transparent sheet tacked at the origin of a Wulff net.

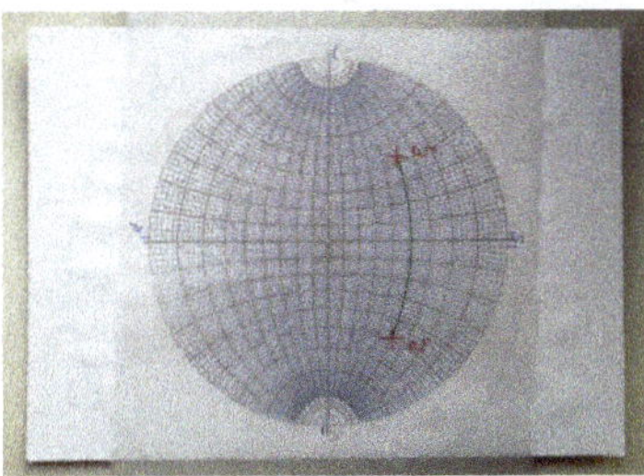

The transparent sheet is rotated and the central angle is read along the common meridian to both points P_1 and P_2.

Other Formulations and Generalizations

Some authors define stereographic projection from the north pole (0, 0, 1) onto the plane $z = -1$, which is tangent to the unit sphere at the south pole (0, 0, −1). The values X and Y produced by this projection are exactly twice those produced by the equatorial projection described in the preceding section. For example, this projection sends the equator to the circle of radius 2 centered at the origin. While the equatorial projection produces no infinitesimal area distortion along the equator, this pole-tangent projection instead produces no infinitesimal area distortion at the south pole.

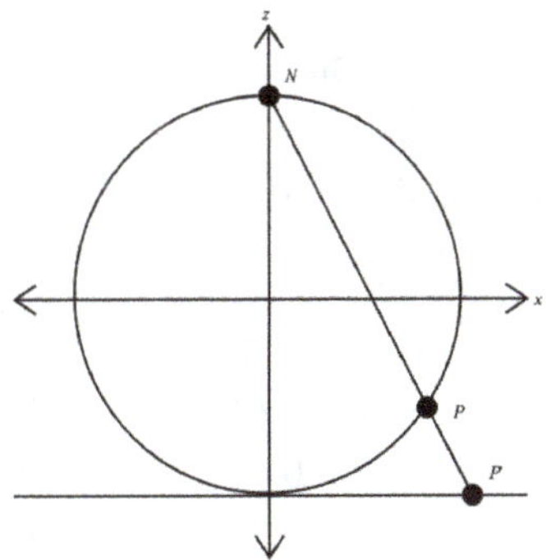

Stereographic projection of the unit sphere from the north pole onto the plane $z = -1$, shown here in cross section

Other authors use a sphere of radius 1/2 and the plane $z = -1/2$. In this case the formulae become

$$(x, y, z) \rightarrow (\xi, \eta) = \left(\frac{x}{\frac{1}{2} - z}, \frac{y}{\frac{1}{2} - z} \right),$$

$$(\xi, \eta) \rightarrow (x, y, z) = \left(\frac{\xi}{1 + \xi^2 + \eta^2}, \frac{\eta}{1 + \xi^2 + \eta^2}, \frac{-1 + \xi^2 + \eta^2}{2 + 2\xi^2 + 2\eta^2} \right).$$

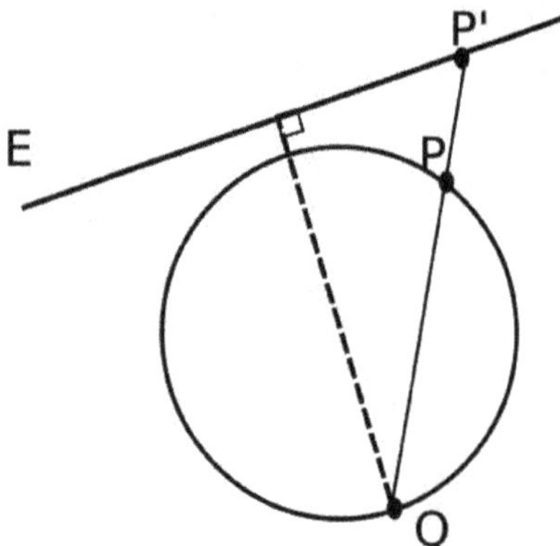

Stereographic projection of a sphere from a point Q onto the plane E, shown here in cross section

In general, one can define a stereographic projection from any point Q on the sphere onto any plane E such that

- E is perpendicular to the diameter through Q, and
- E does not contain Q.

As long as E meets these conditions, then for any point P other than Q the line through P and Q meets E in exactly one point P', which is defined to be the stereographic projection of P onto E.

All of the formulations of stereographic projection described thus far have the same essential properties. They are smooth bijections (diffeomorphisms) defined everywhere except at the projection point. They are conformal and not area-preserving.

More generally, stereographic projection may be applied to the n-sphere S^n in $(n + 1)$-dimensional Euclidean space E^{n+1}. If Q is a point of S^n and E a hyperplane in E^{n+1}, then the stereographic projection of a point $P \in S^n - \{Q\}$ is the point P' of intersection of the line QP with E. In Cartesian coordinates (x_i, i from 0 to n) on the sphere and (X_i, i from 1 to n) on the plane, the projection from $Q = (1, 0, 0, \ldots, 0)$ is given by

$$X_i = \frac{x_i}{1 - x_0} \quad (i \text{ from } 1 \text{ to } n)$$

Defining

$$S^2 = \sum_{j=1}^{n} X_j^2$$

the inverse is given by

$$x_0 = \frac{S^2 - 1}{S^2 + 1} \quad \text{and} \quad x_i = \frac{2X_i}{S^2 + 1} \quad (i \text{ from1 to } n)$$

Still more generally, suppose that S is a (nonsingular) quadric hypersurface in the projective space P^{n+1}. In other words, S is the locus of zeros of a non-singular quadratic form $f(x_0, \ldots, x_{n+1})$ in the homogeneous coordinates x_i. Fix any point Q on S and a hyperplane E in P^{n+1} not containing Q. Then the stereographic projection of a point P in $S - \{Q\}$ is the unique point of intersection of QP with E. As before, the stereographic projection is conformal and invertible outside of a "small" set. The stereographic projection presents the quadric hypersurface as a rational hypersurface. This construction plays a role in algebraic geometry and conformal geometry.

Applications within Mathematics

Complex Analysis

Although any stereographic projection misses one point on the sphere (the projection point), the entire sphere can be mapped using two projections from distinct projection points. In other words, the sphere can be covered by two stereographic parametrizations (the inverses of the projections) from the plane. The parametrizations can be chosen to induce the same orientation on the sphere. Together, they describe the sphere as an oriented surface (or two-dimensional manifold).

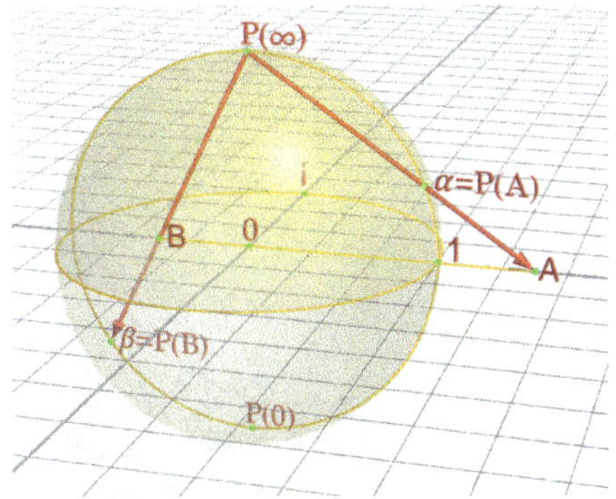

The complex plane and the Riemann sphere above it

This construction has special significance in complex analysis. The point (X, Y) in the real plane can be identified with the complex number $\zeta = X + iY$. The stereographic projection from the north pole onto the equatorial plane is then

$$\zeta = \frac{x+iy}{1-z},$$

$$(x,y,z) = \left(\frac{2\operatorname{Re}\zeta}{1+\bar{\zeta}\zeta}, \frac{2\operatorname{Im}\zeta}{1+\bar{\zeta}\zeta}, \frac{-1+\bar{\zeta}\zeta}{1+\bar{\zeta}\zeta}\right).$$

Similarly, letting $\xi = X - iY$ be another complex coordinate, the functions

$$\xi = \frac{x-iy}{1+z},$$

$$(x,y,z) = \left(\frac{2\operatorname{Re}\xi}{1+\bar{\xi}\xi}, \frac{-2\operatorname{Im}\xi}{1+\bar{\xi}\xi}, \frac{1-\bar{\xi}\xi}{1+\bar{\xi}\xi}\right).$$

define a stereographic projection from the south pole onto the equatorial plane. The transition maps between the ζ- and ξ-coordinates are then $\zeta = 1/\xi$ and $\xi = 1/\zeta$, with ζ approaching 0 as ξ goes to infinity, and *vice versa*. This facilitates an elegant and useful notion of infinity for the complex numbers and indeed an entire theory of meromorphic functions mapping to the Riemann sphere. The standard metric on the unit sphere agrees with the Fubini–Study metric on the Riemann sphere.

Visualization of Lines and Planes

The set of all lines through the origin in three-dimensional space forms a space called the real projective plane. This space is difficult to visualize, because it cannot be embedded in three-dimensional space.

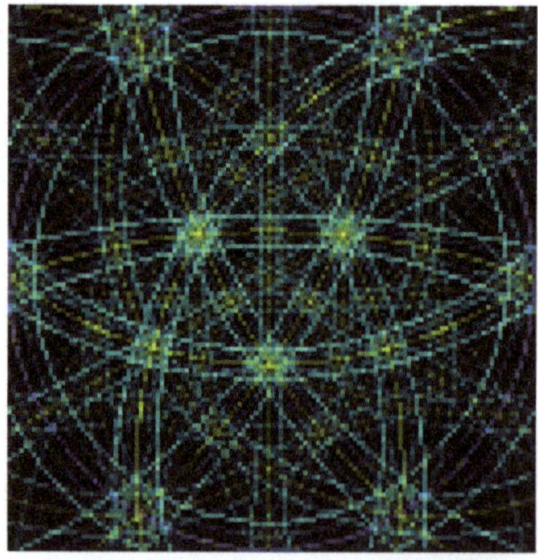

Animation of Kikuchi lines of four of the eight <111> zones in an fcc crystal. Planes edge-on (banded lines) intersect at fixed angles.

However, one can "almost" visualize it as a disk, as follows. Any line through the origin intersects the southern hemisphere $z \leq 0$ in a point, which can then be stereographically projected to a point on a disk. Horizontal lines intersect the southern hemisphere in two antipodal points along the equator, either of which can be projected to the disk; it is understood that antipodal points on the boundary of the disk represent a single line. So any set of lines through the origin can be pictured, almost perfectly, as a set of points in a disk.

Also, every plane through the origin intersects the unit sphere in a great circle, called the *trace* of the plane. This circle maps to a circle under stereographic projection. So the projection lets us visualize planes as circular arcs in the disk. Prior to the availability of computers, stereographic projections with great circles often involved drawing large-radius arcs that required use of a beam compass. Computers now make this task much easier.

Further associated with each plane is a unique line, called the plane's *pole*, that passes through the origin and is perpendicular to the plane. This line can be plotted as a point on the disk just as any line through the origin can. So the stereographic projection also lets us visualize planes as points in the disk. For plots involving many planes, plotting their poles produces a less-cluttered picture than plotting their traces.

This construction is used to visualize directional data in crystallography and geology, as described below.

Other Visualization

Stereographic projection is also applied to the visualization of polytopes. In a Schlegel diagram, an n-dimensional polytope in R^{n+1} is projected onto an n-dimensional sphere, which is then stereographically projected onto R^n. The reduction from R^{n+1} to R^n can make the polytope easier to visualize and understand.

Arithmetic Geometry

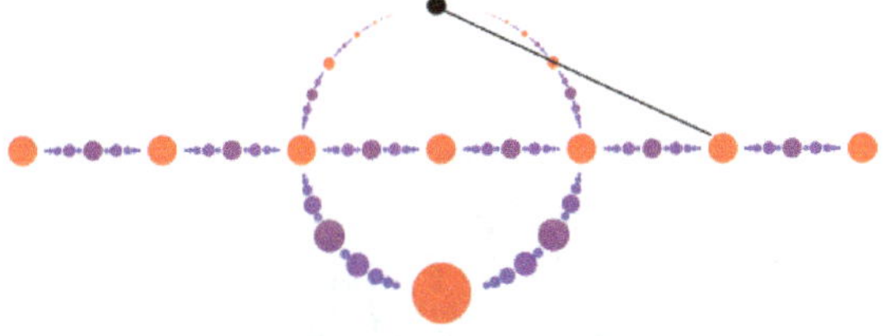

The rational points on a circle correspond, under stereographic projection, to the rational points of the line.

In elementary arithmetic geometry, stereographic projection from the unit circle provides a means to describe all primitive Pythagorean triples. Specifically, stereographic projection from the north pole (0,1) onto the x-axis gives a one-to-one correspondence between the rational number points (x, y) on the unit circle (with $y \neq 1$) and the rational points of the x-axis. If $(m/n, 0)$ is a rational point on the x-axis, then its inverse stereographic projection is the point

$$\left(\frac{2mn}{n^2+m^2}, \frac{n^2-m^2}{n^2+m^2}\right)$$

which gives Euclid's formula for a Pythagorean triple.

Tangent Half-angle Substitution

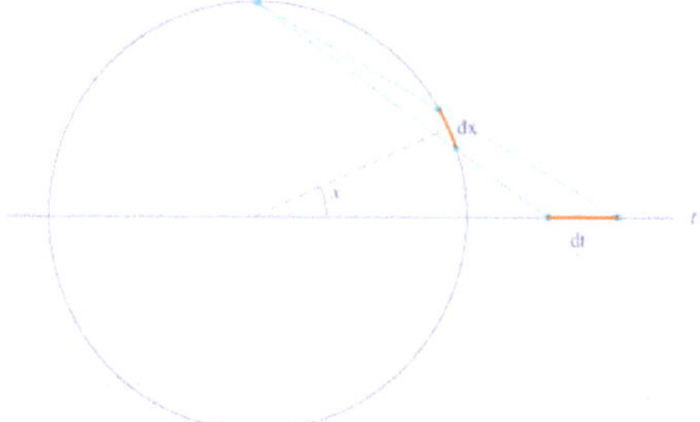

The pair of trigonometric functions $(\sin x, \cos x)$ can be thought of as parametrizing the unit circle. The stereographic projection gives an alternative parametrization of the unit circle:

$$\cos x = \frac{t^2-1}{t^2+1}, \quad \sin x = \frac{2t}{t^2+1}.$$

Under this reparametrization, the length element dx of the unit circle goes over to

$$dx = \frac{2dt}{t^2+1}.$$

This substitution can sometimes simplify integrals involving trigonometric functions.

Applications to other Disciplines

Cartography

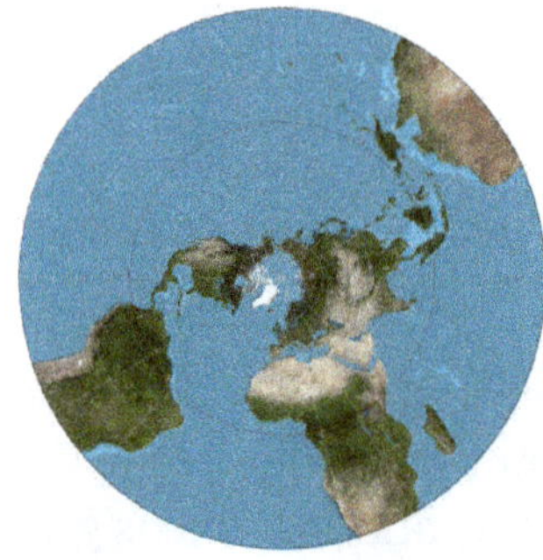

Stereographic projection of the world north of 30°S. 15° graticule.

Rumold Mercator's map

Joan Blaeu's map

The fundamental problem of cartography is that no map from the sphere to the plane can accurately represent both angles and areas. In general, area-preserving map projections are preferred for statistical applications, while angle-preserving (conformal) map projections are preferred for navigation.

Stereographic projection falls into the second category. When the projection is centered at the Earth's north or south pole, it has additional desirable properties: It sends meridians to rays emanating from the origin and parallels to circles centered at the origin.

The stereographic is the only projection that maps all circles of a sphere to circles. This property is valuable in planetary mapping when craters are typical features. The set of circles passing through the point of projection have unbounded radius, and therefore degenerate into lines.

Crystallography

In crystallography, the orientations of crystal axes and faces in three-dimensional space are a central geometric concern, for example in the interpretation of X-ray and electron diffraction patterns. These orientations can be visualized as in the section Visualization of lines and planes above. That is, crystal axes and poles to crystal planes are intersected with the northern hemisphere and then plotted using stereographic projection. A plot of poles is called a pole figure.

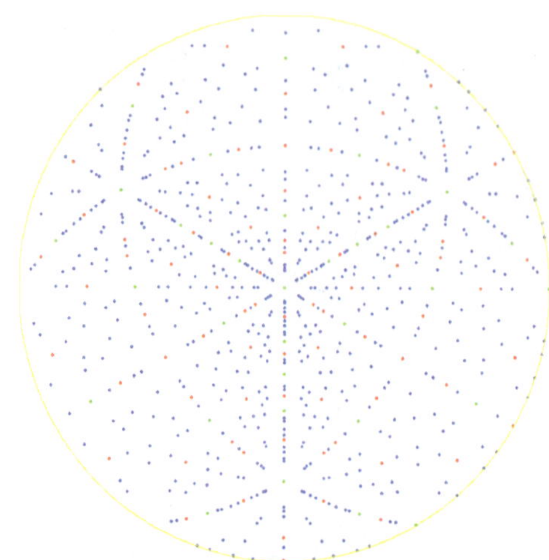

A crystallographic pole figure for the diamond lattice in direction

In electron diffraction, Kikuchi line pairs appear as bands decorating the intersection between lattice plane traces and the Ewald sphere thus providing *experimental access* to a crystal's stereographic projection. Model Kikuchi maps in reciprocal space, and fringe visibility maps for use with bend contours in direct space, thus act as road maps for exploring orientation space with crystals in the transmission electron microscope.

Geology

Researchers in structural geology are concerned with the orientations of planes and lines for a number of reasons. The foliation of a rock is a planar feature that often contains a linear feature called lineation. Similarly, a fault plane is a planar feature that may contain linear features such as slickensides.

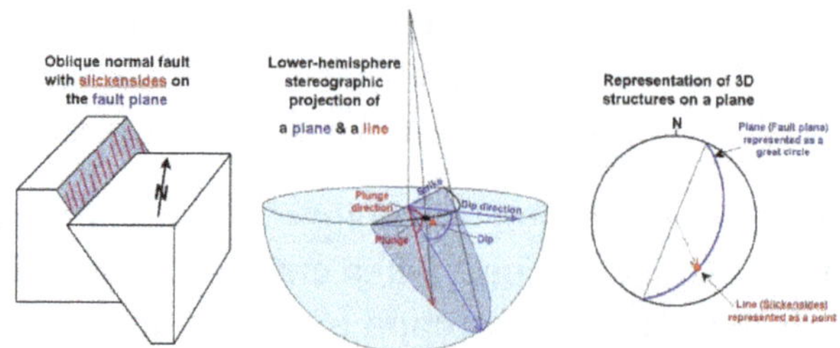

Use of lower hemisphere stereographic projection to plot planar and linear data in structural geology, using the example of a fault plane with a slickenside lineation

These orientations of lines and planes at various scales can be plotted using the methods of the Visualization of lines and planes section above. As in crystallography, planes are typically plotted by their poles. Unlike crystallography, the southern hemisphere is used instead of the northern one (because the geological features in question lie below the Earth's surface). In this context the stereographic projection is often referred to as the equal-angle lower-hemisphere projection. The equal-area lower-hemisphere projection defined by the

Lambert azimuthal equal-area projection is also used, especially when the plot is to be subjected to subsequent statistical analysis such as density contouring.

Photography

Stereographic projection of the spherical panorama of the Last Supper sculpture by Michele Vedani in Esino Lario, Lombardy, Italy during Wikimania 2016

Some fisheye lenses use a stereographic projection to capture a wide-angle view. Compared to more traditional fisheye lenses which use an equal-area projection, areas close to the edge retain their shape, and straight lines are less curved. However, stereographic fisheye lenses are typically more expensive to manufacture. Image remapping software, such as Panotools, allows the automatic remapping of photos from an equal-area fisheye to a stereographic projection.

The stereographic projection has been used to map spherical panoramas. This results in effects known as a *little planet* (when the center of projection is the nadir) and a *tube* (when the center of projection is the zenith).

The popularity of using stereographic projections to map panoramas over other azimuthal projections is attributed to the shape preservation that results from the conformality of the projection.

References

- Evans, James (1998), The History and Practice of Ancient Astronomy, Oxford: Oxford University Press, pp. 102–103, ISBN 9780199874453.
- Noordung, Hermann; et al. (1995) [1929]. The Problem With Space Travel. Translation from original German. DIANE Publishing. p. 72. ISBN 978-0-7881-1849-4.
- Hanspeter Schaub, John L. Junkins (2003). "Rigid body kinematics". Analytical mechanics of space systems. American Institute of Aeronautics and Astronautics. p. 71. ISBN 1-56347-563-4.

- Snyder, John P. (1993). Flattening the earth: two thousand years of map projections. University of Chicago Press. ISBN 0-226-76746-9.
- Choosing a World Map. Falls Church, Virginia: American Congress on Surveying and Mapping. 1988. p. 1. ISBN 0-9613459-2-6.
- Olivares, Miriam. "Geographic Information Systems at Yale: Geocoding Resources". guides.library.yale.edu. Retrieved 2016-06-22.
- Kwok, Geodetic Survey Section Lands Department Hong Kong. "Geodetic Datum Transformation, p.24" (PDF). Geodetic Survey Section Lands Department Hong Kong. Retrieved 4 March 2014.
- Osborne, Peter (2013). "Chapters 5,6". The Mercator Projections. doi:10.5281/zenodo.35392. doi:10.5281/zenodo.35562 for LaTeX code and figures.
- A geostationary Earth orbit satellite model using Easy Java Simulation Loo Kang Wee and Giam Hwee Goh 2013 Phys. Educ. 48 72
- Karney, C. F. F. (2013). "Algorithms for geodesics". Journal of Geodesy. 87 (1): 43–42. arXiv:1109.4448. Bibcode:2013JGeod..87...43K. doi:10.1007/s00190-012-0578-z Addenda.

5

Applications of Geographic Information Systems

The applications of geographic information system explained in the chapter are GIS applications, satellite imagery, crime mapping, Google Earth, map algebra, map regression etc. The software and hardware systems that enable users to capture, store and manage geographic data are known as geographic information systems. The text strategically encompasses and incorporates the major application of geographic information system.

GIS Applications

Geographic information systems (GIS) (also known as Geospatial information systems) are computer software and hardware systems that enable users to capture, store, analyse and manage spatially referenced data. GISs have transformed the way spatial (geographic) data, relationships and patterns in the world are able to be interactively queried, processed, analysed, mapped, modelled, visualised, and displayed for an increasingly large range of users, for a multitude of purposes.

Examples of GIS Applications

Uses of GIS range from indigenous people, communities, research institutions, environmental scientists, health organisations, land use planners, businesses, and government agencies at all levels.

Some examples include:

- Crime mapping
- Historical geographic information systems
- GIS and Hydrology
- Remote sensing applications
- Traditional knowledge gis
- Public Participation GIS
- Road networking

- Wastewater and stormwater systems
- Waste management

Satellite Imagery

The first images from space were taken on the sub-orbital V-2 rocket flight launched by the U.S. on October 24, 1946.

Satellite imagery consists of images of Earth or other planets collected by satellites. Imaging satellites are operated by governments and businesses around the world. Satellite imaging companies sell images under licence. Images are licensed to governments and businesses such as Apple Maps and Google Maps.

History

The satellite images were made from pixels. The first crude image taken by the satellite Explorer 6 shows a sunlit area of the Central Pacific Ocean and its cloud cover. The photo was taken when the satellite was about 17,000 mi (27,000 km) above the surface of the earth on August 14, 1959. At the time, the satellite was crossing Mexico.

The first images from space were taken on sub-orbital flights. The U.S-launched V-2 flight on October 24, 1946 took one image every 1.5 seconds. With an apogee of 65 miles (105 km), these photos were from five times higher than the previous record, the 13.7 miles (22 km) by the Explorer II balloon mission in 1935. The first satellite (orbital) photographs of Earth were made on August 14, 1959 by the U.S. Explorer 6. The first satellite photographs of the Moon might have been made on October 6, 1959 by the Soviet satellite Luna 3, on a mission to photograph the far side of the Moon. The Blue Marble photograph was taken from space in 1972, and has become very popular in the media and among the public. Also in 1972 the United States started the Landsat program, the largest program for acquisition of imagery of Earth from space. Landsat Data Continuity Mission, the most recent Landsat satellite, was launched on 11 February 2013. In 1977, the first real time satellite imagery was acquired by the USA's KH-11 satellite system.

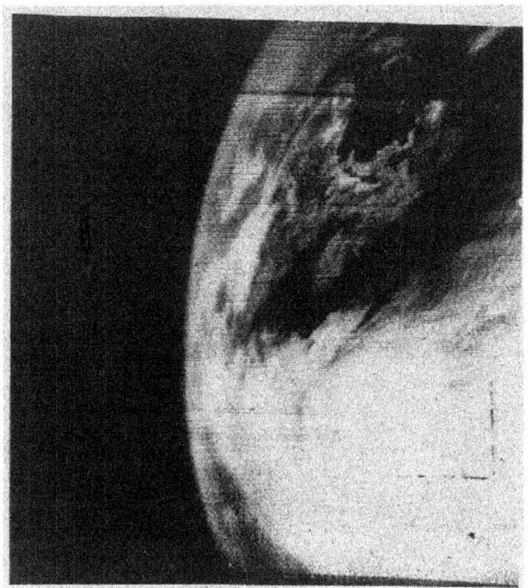

The first television image of Earth from space transmitted by the TIROS-1 weatherA satellite in 1960.

All satellite images produced by NASA are published by NASA Earth Observatory and are freely available to the public. Several other countries have satellite imaging programs, and a collaborative European effort launched the ERS and Envisat satellites carrying various sensors. There are also private companies that provide commercial satellite imagery. In the early 21st century satellite imagery became widely available when affordable, easy to use software with access to satellite imagery databases was offered by several companies and organizations.

Uses

Satellite images have many applications in meteorology, oceanography, fishing, agriculture, biodiversity conservation, forestry, landscape, geology, cartography, regional planning, education, intelligence and warfare. Images can be in visible colours and in other spectra. There are also elevation maps, usually made by radar images. Interpretation and analysis of satellite imagery is conducted using specialized remote sensing applications.

Satellite photography can be used to produce composite images of an entire hemisphere

...or to map a small area of the Earth, such as this photo of the countryside of Haskell County, Kansas, United States.

Resolution and Data

There are four types of resolution when discussing satellite imagery in remote sensing: spatial, spectral, temporal, and radiometric. Campbell (2002) defines these as follows:

- spatial resolution is defined as the pixel size of an image representing the size of the surface area (i.e. m^2) being measured on the ground, determined by the sensors' instantaneous field of view (IFOV);
- spectral resolution is defined by the wavelength interval size (discreet segment of the Electromagnetic Spectrum) and number of intervals that the sensor is measuring;
- temporal resolution is defined by the amount of time (e.g. days) that passes between imagery collection periods for a given surface location
- Radiometric resolution is defined as the ability of an imaging system to record many levels of brightness (contrast for example) and to the effective bit-depth

of the sensor (number of grayscale levels) and is typically expressed as 8-bit (0-255), 11-bit (0-2047), 12-bit (0-4095) or 16-bit (0-65,535).

- Geometric resolution refers to the satellite sensor's ability to effectively image a portion of the Earth's surface in a single pixel and is typically expressed in terms of Ground sample distance, or GSD. GSD is a term containing the overall optical and systemic noise sources and is useful for comparing how well one sensor can "see" an object on the ground within a single pixel. For example, the GSD of Landsat is ~30m, which means the smallest unit that maps to a single pixel within an image is ~30m x 30m. The latest commercial satellite (GeoEye 1) has a GSD of 0.41 m. This compares to a 0.3 m resolution obtained by some early military film based Reconnaissance satellite such as Corona.

The resolution of satellite images varies depending on the instrument used and the altitude of the satellite's orbit. For example, the Landsat archive offers repeated imagery at 30 meter resolution for the planet, but most of it has not been processed from the raw data. Landsat 7 has an average return period of 16 days. For many smaller areas, images with resolution as high as 41 cm can be available.

Satellite imagery is sometimes supplemented with aerial photography, which has higher resolution, but is more expensive per square meter. Satellite imagery can be combined with vector or raster data in a GIS provided that the imagery has been spatially rectified so that it will properly align with other data sets.

Imaging Satellites

GeoEye

GeoEye's GeoEye-1 satellite was launched September 6, 2008. The GeoEye-1 satellite has the high resolution imaging system and is able to collect images with a ground resolution of 0.41 meters (16 inches) in the panchromatic or black and white mode. It collects multispectral or color imagery at 1.65-meter resolution or about 64 inches.

DigitalGlobe

DigitalGlobe's WorldView-2 satellite provides high resolution commercial satellite imagery with 0.46 m spatial resolution (panchromatic only). The 0.46 meters resolution of WorldView-2's panchromatic images allows the satellite to distinguish between objects on the ground that are at least 46 cm apart. Similarly DigitalGlobe's QuickBird satellite provides 0.6 meter resolution (at NADIR) panchromatic images.

DigitalGlobe's WorldView-3 satellite provides high resolution commercial satellite imagery with 0.31 m spatial resolution. WVIII also carries a short wave infrared sensor and an atmospheric sensor

Spot Image

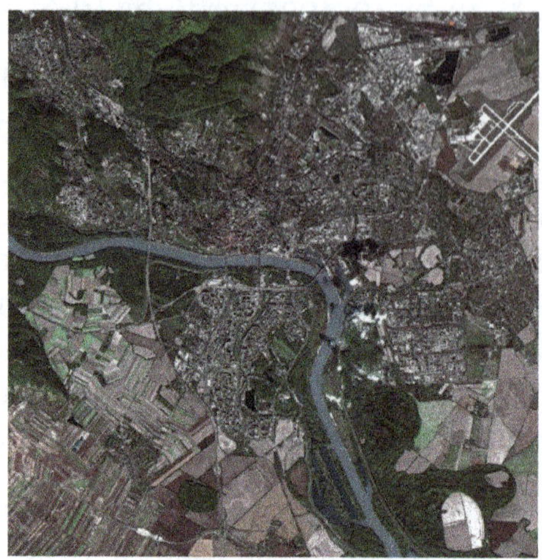

SPOT image of Bratislava

The 3 SPOT satellites in orbit (Spot 2, 4 and 5) provide images with a large choice of resolutions – from 2.5 m to 1 km. Spot Image also distributes multiresolution data from other optical satellites, in particular from Formosat-2 (Taiwan) and Kompsat-2 (South Korea) and from radar satellites (TerraSar-X, ERS, Envisat, Radarsat). Spot Image will also be the exclusive distributor of data from the forthcoming very-high resolution Pleiades satellites with a resolution of 0.50 meter or about 20 inches. The first launch is planned for the end of 2011. The company also offers infrastructures for receiving and processing, as well as added value options.

ASTER

The Advanced Spaceborne Thermal Emission and Reflection Radiometer (ASTER) is an imaging instrument onboard Terra, the flagship satellite of NASA's Earth Observing System (EOS) launched in December 1999. ASTER is a cooperative effort between NASA, Japan's Ministry of Economy, Trade and Industry (METI), and Japan Space Systems (J-spacesystems). ASTER data is used to create detailed maps of land surface temperature, reflectance, and elevation. The coordinated system of EOS satellites, including Terra, is a major component of NASA's Science Mission Directorate and the Earth Science Division. The goal of NASA Earth Science is to develop a scientific understanding of the Earth as an integrated system, its response to change, and to better predict variability and trends in climate, weather, and natural hazards.

- Land surface climatology—investigation of land surface parameters, surface temperature, etc., to understand land-surface interaction and energy and moisture fluxes
- Vegetation and ecosystem dynamics—investigations of vegetation and soil dis-

tribution and their changes to estimate biological productivity, understand land-atmosphere interactions, and detect ecosystem change

- Volcano monitoring—monitoring of eruptions and precursor events, such as gas emissions, eruption plumes, development of lava lakes, eruptive history and eruptive potential
- Hazard monitoring—observation of the extent and effects of wildfires, flooding, coastal erosion, earthquake damage, and tsunami damage
- Hydrology—understanding global energy and hydrologic processes and their relationship to global change; included is evapotranspiration from plants
- Geology and soils—the detailed composition and geomorphologic mapping of surface soils and bedrocks to study land surface processes and earth's history
- Land surface and land cover change—monitoring desertification, deforestation, and urbanization; providing data for conservation managers to monitor protected areas, national parks, and wilderness areas

BlackBridge

BlackBridge, previously known as RapidEye, operates a constellation of five satellites, launched in August 2008, the RapidEye constillation contains identical multispectral sensors which are equally calibrated. Therefore, an image from one satellite will be equivalent to an image from any of the other four, allowing for a large amount of imagery to be collected (4 million km^2 per day), and daily revisit to an area. Each travel on the same orbital plane at 630 km, and deliver images in 5 meter pixel size. RapidEye satellite imagery is especially suited for agricultural, environmental, cartographic and disaster management applications. The company not only offers their imagery, but consults with their customers to create services and solutions based on analysis of this imagery .

ImageSat International

Earth Resource Observation Satellites, better known as "EROS" satellites, are lightweight, low earth orbiting, high-resolution satellites designed for fast maneuvering between imaging targets. In the commercial high-resolution satellite market, EROS is the smallest very high resolution satellite; it is very agile and thus enables very high performances. The satellites are deployed in a circular sun-synchronous near polar orbit at an altitude of 510 km (+/- 40 km). EROS satellites imagery applications are primarily for intelligence, homeland security and national development purposes but also employed in a wide range of civilian applications, including: mapping, border control, infrastructure planning, agricultural monitoring, environmental monitoring, disaster response, training and simulations, etc.

EROS A – a high resolution satellite with 1.9-1.2m resolution panchromatic was launched on December 5, 2000.

EROS B - the second generation of Very High Resolution satellites with 70 cm resolution panchromatic, was launched on April 25, 2006.

Meteosat

Model of a first generation Meteosat geostationary satellite.

The Meteosat-2 geostationary weather satellite began operationally to supply imager data on 16 August 1981. Eumetsat has operated the Meteosats since 1987.

- The *Meteosat visible and infrared imager (MVIRI)*, three-channel imager: visible, infrared and water vapour; It operates on the first generation Meteosat, Meteosat-7 being still active.
- The 12-channel *Spinning Enhanced Visible and Infrared Imager (SEVIRI)* includes similar channels to those used by MVIRI, providing continuity in climate data over three decades; Meteosat Second Generation (MSG).
- The *Flexible Combined Imager (FCI)* on Meteosat Third Generation (MTG) will also include similar channels, meaning that all three generations will have provided over 60 years of climate data.

Disadvantages

Because the total area of the land on Earth is so large and because resolution is relatively high, satellite databases are huge and image processing (creating useful images from the raw data) is time-consuming. Depending on the sensor used, weather conditions can affect image quality: for example, it is difficult to obtain images for areas of frequent cloud cover such as mountain-tops. For such reasons, publicly available satellite image datasets are typically processed for visual or scientific commercial use by third parties.

Commercial satellite companies do not place their imagery into the public domain and do not sell their imagery; instead, one must be licensed to use their imagery. Thus, the ability to legally make derivative products from commercial satellite imagery is minimized.

Privacy concerns have been brought up by some who wish not to have their property shown from above. Google Maps responds to such concerns in their FAQ with the following statement: "*We understand your privacy concerns... The images that Google Maps displays are no different from what can be seen by anyone who flies over or drives by a specific geographic location.*"

Moving Images

In 2005 the Australian company Astrovision (ASX: HZG) announced plans to launch the first commercial geostationary satellite in the Asia-Pacific. It is intended to provide true color, real-time live satellite feeds, with down to 250 metres resolution over the entire Asia-Pacific region, from India to Hawaii and Japan to Australia. They were going to provide this content to users of 3G mobile phones, over Pay TV as a weather channel, and to corporate and government users.

Potential customers were excited by the possibilities offered, but they were unwilling (or, in government cases, generally unable) to sign contracts for a service that would not be delivered for 3–4 years (the length of time required to build and launch the satellite). AstroVision ran low on funds and was forced to shut down the program in 2006.

Crime Mapping

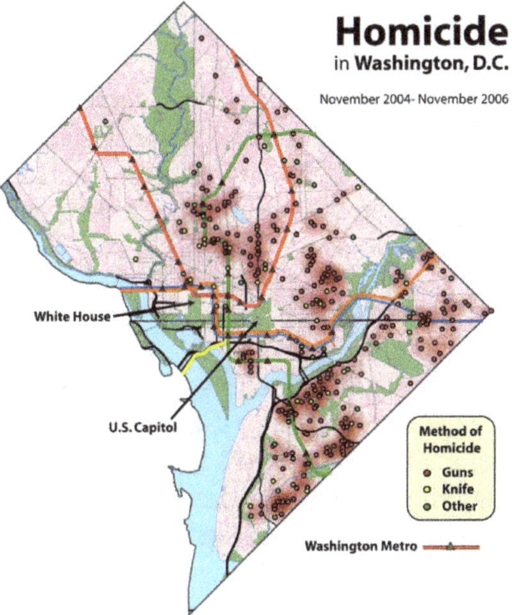

Mapping of homicides in Washington D.C.

Crime mapping is used by analysts in law enforcement agencies to map, visualize, and analyze crime incident patterns. It is a key component of crime analysis and the CompStat policing strategy. Mapping crime, using Geographic Information Systems (GIS), allows crime analysts to identify crime hot spots, along with other trends and patterns.

Overview

Using GIS, crime analysts can overlay other datasets such as census demographics, locations of pawn shops, schools, etc., to better understand the underlying causes of crime and help law enforcement administrators to devise strategies to deal with the problem. GIS is also useful for law enforcement operations, such as allocating police officers and dispatching to emergencies.

Underlying theories that help explain spatial behavior of criminals include environmental criminology, which was devised in the 1980s by Patricia and Paul Brantingham, routine activity theory, developed by Lawrence Cohen and Marcus Felson and originally published in 1979, and rational choice theory, developed by Ronald V. Clarke and Derek Cornish, originally published in 1986. In recent years, crime mapping and analysis has incorporated spatial data analysis techniques that add statistical rigor and address inherent limitations of spatial data, including spatial autocorrelation and spatial heterogeneity. Spatial data analysis helps one analyze crime data and better understand why and not just where crime is occurring.

Research into computer-based crime mapping started in 1986, when the National Institute of Justice (NIJ) funded a project in the Chicago Police Department to explore crime mapping as an adjunct to community policing. That project was carried out by the CPD in conjunction with the Chicago Alliance for Neighborhood Safety, the University of Illinois at Chicago, and Northwestern University, reported on in the book, *Mapping Crime in Its Community Setting: Event Geography Analysis*. The success of this project prompted NIJ to initiate the Drug Market Analysis Program (with the appropriate acronym D-MAP) in five cities, and the techniques these efforts developed led to the spread of crime mapping throughout the US and elsewhere, including the New York City Police Department's CompStat.

Applications

Crime analysts use crime mapping and analysis to help law enforcement management (e.g. the police chief) to make better decisions, target resources, and formulate strategies, as well as for tactical analysis (e.g. crime forecasting, geographic profiling). New York City does this through the CompStat approach, though that way of thinking deals more with the short term. There are other, related approaches with terms including Information-led policing, Intelligence-led policing, Problem-oriented policing, and Community policing. In some law enforcement agencies, crime analysts work in civilian positions, while in other agencies, crime analysts are sworn officers.

From a research and policy perspective, crime mapping is used to understand patterns of incarceration and recidivism, help target resources and programs, evaluate crime prevention or crime reduction programs (e.g. Project Safe Neighborhoods, Weed & Seed and as proposed in *Fixing Broken Windows*), and further understanding of causes of crime.

The boom of internet technologies, particularly web-based geographic information system (GIS) technologies, is opening new opportunities for use of crime mapping to support crime prevention. Research indicates that the functions provided in web-based crime mapping are less than in most traditional crime mapping software. In conclusion, existing works of web-based crime mapping focus on supporting community policing rather than analytical functions such as pattern analysis and prediction.

Google Earth

Google Earth is a virtual globe, map and geographical information program that was originally called EarthViewer 3D created by Keyhole, Inc, a Central Intelligence Agency (CIA) funded company acquired by Google in 2004. It maps the Earth by the superimposition of images obtained from satellite imagery, aerial photography and geographic information system (GIS) onto a 3D globe. It was originally available with three different licenses, but has since been reduced to just two: Google Earth (a free version with limited function) and Google Earth Pro, which is now free (it previously cost $399 a year) and is intended for commercial use. The third original option, Google Earth Plus, has been discontinued.

The product, re-released as Google Earth in 2005, is available for use on personal computers running Windows 2000 and above, Mac OS X 10.3.9 and above, Linux kernel: 2.6 or later (released on June 12, 2006), and FreeBSD. Google Earth is also available as a browser plugin which was released on May 28, 2008. It was also made available for mobile viewers on the iPhone OS on October 28, 2008, as a free download from the App Store, and is available to Android users as a free app in the Google Play store. In addition to releasing an updated Keyhole based client, Google also added the imagery from the Earth database to their web-based mapping software, Google Maps. The release of Google Earth in June 2005 to the public caused a more than tenfold increase in media coverage on virtual globes between 2004 and 2005, driving public interest in geospatial technologies and applications. As of October 2011, Google Earth has been downloaded more than a billion times.

Google Earth displays satellite images of varying resolution of the Earth's surface, allowing users to see things like cities and houses looking perpendicularly down or at an oblique angle. The degree of resolution available is based somewhat on the points of interest and popularity, but most land (except for some islands) is covered in at least 15 meters of resolution. Maps showing a visual representation of Google Earth coverage Melbourne, Victoria, Australia; Las Vegas, Nevada, USA; and Cambridge, Cambridgeshire, United Kingdom include examples of the highest resolution, at 15 cm (6 inches). Google Earth allows users to search for addresses for some countries, enter coordinates, or simply use the mouse to browse to a location.

For large parts of the surface of the Earth only 2D images are available, from almost vertical photography. Viewing this from an oblique angle, there is perspective in the sense that objects which are horizontally far away are seen smaller, like viewing a large photograph, not quite like a 3D view.

For other parts of the surface of the Earth, 3D images of terrain and buildings are available. Google Earth uses digital elevation model (DEM) data collected by NASA's Shuttle Radar Topography Mission (SRTM). This means one can view almost the entire earth in three dimensions. Since November 2006, the 3D views of many mountains, including Mount Everest, have been improved by the use of supplementary DEM data to fill the gaps in SRTM coverage.

Some people use the applications to add their own data, making them available through various sources, such as the Bulletin Board Systems (BBS) or blogs mentioned in the link section below. Google Earth is able to show various kinds of images overlaid on the surface of the earth and is also a Web Map Service client. Google Earth supports managing three-dimensional Geospatial data through Keyhole Markup Language (KML).

Detail

Google Earth is simply based on 3D maps, with the capability to show 3D buildings and structures (such as bridges), which consist of users' submissions using SketchUp, a 3D modeling program software. In prior versions of Google Earth (before Version 4), 3D buildings were limited to a few cities, and had poorer rendering with no textures. Many buildings and structures from around the world now have detailed 3D structures; including (but not limited to) those in the United States, Canada, Mexico, India, Japan, United Kingdom, Spain, Germany, Pakistan and the cities, Amsterdam and Alexandria. In August 2007, Hamburg became the first city entirely shown in 3D, including textures such as façades. The 'Westport3D' model was created by 3D imaging firm AM3TD using long-distance laser scanning technology and digital photography and is the first such model of an Irish town to be created. As it was developed initially to aid Local Government in carrying out their town planning functions it includes the highest-resolution photo-realistic textures to be found anywhere in Google Earth. Three-dimensional renderings are available for certain buildings and structures around the world via Google's 3D Warehouse and other websites. In June 2012, Google announced that it will start to replace user submitted 3D buildings with auto-generated 3D mesh buildings starting with major cities. Although there are many cities on Google Earth that are fully or partially 3D, more are available in the Earth Gallery. The Earth Gallery is a library of modifications of Google Earth people have made. In the library there are not only modifications for 3D buildings, but also models of earthquakes using the Google Earth model, 3D forests, and much more.

Recently, Google added a feature that allows users to monitor traffic speeds at loops located every 200 yards in real-time. In 2007, Google began offering traffic data in real-time, based on information crowdsourced from the GPS-identified locations of cellular phone users. In version 4.3 released on April 15, 2008, Google Street View was fully integrated into the program allowing the program to provide an on the street level view in many locations.

On January 31, 2010, the entirety of Google Earth's ocean floor imagery was updated to new images by SIO, NOAA, US Navy, NGA, and GEBCO. The new images have caused smaller islands, such as some atolls in the Maldives, to be rendered invisible despite their shores being completely outlined.

Uses

Google Earth may be used to perform some day-to-day tasks and for other purposes.

- Google Earth can be used to view areas subjected to widespread disasters if Google supplies up-to-date images. For example, after the January 12, 2010 Haiti earthquake images of Haiti were made available on January 17.
- With Google's push for the inclusion of Google Earth in the Classroom, teachers are adopting Google Earth in the classroom for lesson planning, such as teaching students geographical themes (location, culture, characteristics, human interaction, and movement) to creating mashups with other web applications such as Wikipedia.
- One can explore and place location bookmarks on the Moon and Mars.
- One can get directions using Google Earth, using variables such as street names, cities, and establishments. But the addresses must by typed in search field, one cannot simply click on two spots on the map.
- Google Earth can function as a hub of knowledge, pertaining the users location. By enabling certain options, one can see the location of gas stations, restaurants, museums, and other public establishments in their area. Google Earth can also dot the map with links to images, YouTube videos, and Wikipedia articles relevant to the area being viewed.
- One can create custom image overlays for planning trips, hikes on handheld GPS units.
- Google Earth can be used to map homes and select a random sample for research in developing countries.

All of these features are also released by Google Earth Blog.

Features

Wikipedia and Panoramio Integration

In December 2006, Google Earth added a new layer called "Geographic Web" that includes integration with Wikipedia and Panoramio. In Wikipedia, entries are scraped for coordinates via the Coord templates. There is also a community-layer from the project Wikipedia-World. More coordinates are used, different types are in the display and different languages are supported than the built-in Wikipedia layer. Google announced on May 30, 2007 that it is acquiring Panoramio. In March 2010, Google removed the "Geographic Web" layer. The "Panoramio" layer became part of the main layers and the "Wikipedia" layer was placed in the "More" layer.

Flight Simulator

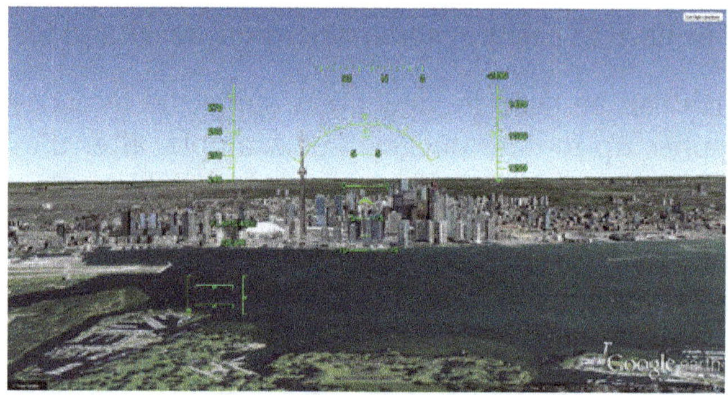

Downtown Toronto as seen from a F-16 Fighting Falcon during a simulated flight

In Google Earth v4.2 a flight simulator was included as a hidden feature. Starting with v4.3 it is no longer hidden. The flight simulator could be accessed by holding down the keys Ctrl, Alt, and A. Initially the F-16 Fighting Falcon and the Cirrus SR-22 were the only aircraft available, and they could be used with only a few airports. However, one can start flight in "current location" and need not to be at an airport. One will face the direction they face when they start the flight simulator. They cannot start flight in ground level view and must be near the ground (approximately 50m-100m above the ground) to start in take-off position. Otherwise they will be in the air with 40% flaps and gears extended (landing position). In addition to keyboard control, the simulator can be controlled with a mouse or joystick. Google Earth v5.1 and higher crashes when starting flight simulator with Saitek and other joysticks. The user can also fly underwater.

Featured Planes

- F-16 Fighting Falcon – A much higher speed and maximum altitude than the Cirrus SR-22, it has the ability to fly at a maximum speed of Mach 2, although a maximum speed of 1678 knots (3108 km/h) can be achieved. The take-off speed is 225 knots, the landing speed is 200 knots (370 km/h).

- Cirrus SR-22 – Although slower and with a lower maximum altitude, the SR-22 is much easier to handle and is preferred for up-close viewing of Google Earth›s imagery. The take-off speed is 75 knots (139 km/h), the landing speed is 70 knots (130 km/h)

The flight simulator can be commanded with the keyboard, mouse or plugged-in joystick. Broadband connection and a high speed computer provides a very realistic experience. The simulator also runs with animation, allowing objects (for example: planes) to animate while on the simulator. Programming language can also be used to make it look like the cockpit of a plane, or for instrument landing.

Sky Mode

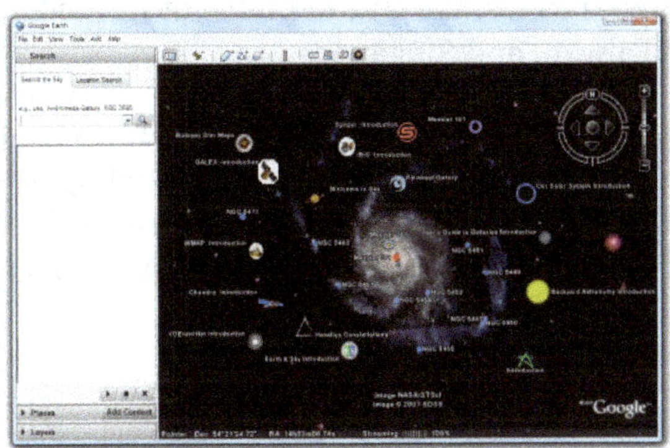

Google Earth in Sky Viewing Mode

Google Sky is a feature that was introduced in Google Earth 4.2 on August 22, 2007, and allows users to view stars and other celestial bodies. It was produced by Google through a partnership with the Space Telescope Science Institute (STScI) in Baltimore, the science operations center for the Hubble Space Telescope. Dr. Alberto Conti and his co-developer Dr. Carol Christian of STScI plan to add the public images from 2007, as well as color images of all of the archived data from Hubble's Advanced Camera for Surveys. Newly released Hubble pictures will be added to the Google Sky program as soon as they are issued. New features such as multi-wavelength data, positions of major satellites and their orbits as well as educational resources will be provided to the Google Earth community and also through Christian and Conti's website for Sky. Also visible on Sky mode are constellations, stars, galaxies and animations depicting the planets in their orbits. A real-time Google Sky mashup of recent astronomical transients, using the VOEvent protocol, is being provided by the VOEventNet collaboration. Google's Earth maps are being updated each 5 minutes.

Google Sky faces competition from Microsoft WorldWide Telescope (which runs only under the Microsoft Windows operating systems) and from Stellarium, a free open source planetarium that runs under Microsoft Windows, OS X, and Linux.

On March 13, 2008, Google made a web-based version of Google Sky available via the internet.

Street View

On April 15, 2008 with version 4.3, Google fully integrated its Street View into Google Earth. In version 6.0, the photo zooming function has been removed because it is incompatible with the new 'seamless' navigation.

Google Street View provides 360° panoramic street-level views and allows users to view parts of selected cities and their surrounding metropolitan areas at ground level. When it was launched on May 25, 2007 for Google Maps, only five cities were included. It has since expanded to more than 40 U.S. cities, and includes the suburbs of many, and in some cases, other nearby cities. Recent updates have now implemented Street View in most of the major cities of Canada, Mexico, Denmark, South Africa, Japan, Spain, Norway, Finland, Sweden, France, the UK, Republic of Ireland, the Netherlands, Italy, Switzerland, Portugal, Taiwan, and Singapore.

Google Street View, when operated, displays photos that were previously taken by a camera mounted on an automobile, and can be navigated by using the mouse to click on photograph icons displayed on the screen in the user's direction of travel. Using these devices, the photos can be viewed in different sizes, from any direction, and from a variety of angles.

Water and Ocean

Introduced in version 5.0 (February 2009), the *Google Ocean* feature allows users to zoom below the surface of the ocean and view the 3D bathymetry beneath the waves. Supporting over 20 content layers, it contains information from leading scientists and oceanographers. On April 14, 2009, Google added underwater terrain data for the Great Lakes. In 2010, Google added underwater terrain data for Lake Baikal.

In June 2011, higher resolution of some deep ocean floor areas increased in focus from 1-kilometer grids to 100 meters thanks to a new synthesis of seafloor topography released through Google Earth. The high-resolution features were developed by oceanographers at Columbia University's Lamont-Doherty Earth Observatory from scientific data collected on research cruises. The sharper focus is available for about 5 percent of the oceans (an area larger than North America). Underwater scenery can be seen of the Hudson Canyon off New York City, the Wini Seamount near Hawaii, and the sharp-edged 10,000-foot-high Mendocino Ridge off the U.S Pacific Coast. There is a Google 2011 Seafloor Tour for those interested in viewing ocean deep terrain.

Historical Imagery

Introduced in version 5.0, Historical Imagery allows users to traverse back in time and study earlier stages of any place. This feature allows research that require analysis of past records of various places.

Mars

A picture of Mars' landscape.

Google Earth 5 includes a separate globe of the planet Mars, that can be viewed and analysed for research purposes. The maps are of a much higher resolution than those on the browser version of Google Mars and it also includes 3D renderings of the Martian terrain. There are also some extremely-high-resolution images from the Mars Reconnaissance Orbiter's HiRISE camera that are of a similar resolution to those of the cities on Earth. Finally, there are many high-resolution panoramic images from various Mars landers, such as the Mars Exploration Rovers, Spirit and Opportunity, that can be viewed in a similar way to Google Street View. Interestingly enough, layers on Google Earth (such as World Population Density) can also be applied to Mars. Layers of Mars can also be applied onto Earth. Mars also has a small application found near the face on Mars. It is called Meliza, and features a chat between the user and an automatic robot speaker. It is useful for research on Mars, but is not recommended for normal conversations.

Moon

One of the lunar landers viewed in Google Moon

On July 20, 2009, the 40th anniversary of the Apollo 11 mission, Google introduced the Google Earth version of Google Moon, which allows users to view satellite images of the Moon. It was announced and demonstrated to a group of invited guests by Google along with Buzz Aldrin at the Newseum in Washington, D.C.

Google Earth Engine

Google Earth Engine is a separate product, not a part of Google Earth.

Liquid Galaxy

Liquid Galaxy is a cluster of computers running Google Earth creating an immersive experience. On September 30, 2010, Google made the configuration and schematics for their rigs public, placing code and setup guides on the Liquid Galaxy wiki.

Liquid Galaxy has also been used as a panoramic photo viewer using KRpano, as well as a Google Street View viewer using Peruse-a-Rue Peruse-a-Rue is a method for synchronizing multiple Maps API clients.

Influences

Google Earth can be traced directly back to a small company named Autometric, now a part of Boeing. A team at Autometric, led by Robert Cowling, created a visualization product named Edge Whole Earth. Bob demonstrated Edge to Michael T. Jones, Chris Tanner and others at SGC in 1996. Several other visualization products using imagery existed at the time, including Performer-based ones, but Michael T. Jones stated emphatically that he had "never thought of the complexities of rendering an entire globe ..." The catch phrase "from outer space to in your face" was coined by Autometric President Dan Gordon, and used to explain his concept for personal/local/global range. Edge blazed a trail as well in broadcasting, being used in 1997 on CBS News with Dan Rather, in print for rendering large images draped over terrain for National Geographic, and used for special effects in the feature film *Shadow Conspiracy* in 1997.

Gordon was a huge fan of the 'Earth' program described in Neal Stephenson's sci-fi classic *Snow Crash*. Indeed, a Google Earth co-founder claimed that Google Earth was modeled after *Snow Crash*, while another co-founder said it was inspired by the short science education film *Powers of Ten*. In fact Google Earth was at least partly inspired by a Silicon Graphics demo called "From Outer Space to in Your Face" which zoomed from space into the Swiss Alps then into the Matterhorn. This launch demo was hosted by an Onyx 3000 with InfiniteReality4 graphics, which supported Clip Mapping and was inspired by the hardware texture paging capability (although it did not use the Clip Mapping) and "Powers of Ten". The first Google Earth implementation called Earth Viewer emerged from Intrinsic Graphics as a demonstration of Chris Tanner's software based implementation of a Clip Mapping texture paging system and was spun off as Keyhole Inc.

Versions and Variations

Mac Version

Since version 4.1.7076.4558 (released on May 9, 2007) onward OS X users can, among other new features, upgrade to the "Plus" version via an option in the Google Earth menu. Some users reported difficulties with Google Earth crashing in the then current version when zooming in. Version 5 of Google Earth for Mac was released in 2009, and version 7 was released concomitantly with the Mac and PC versions on 31 October 2012.

Linux Version

Starting with the version 4 beta Google Earth functions under Linux, as a native port using the Qt toolkit. The Free Software Foundation consider the development of a free compatible client for Google Earth to be a High Priority Free Software Project.

Android Version

Google Earth running on Android

An Android version was released on Monday, February 22, 2010.

iOS Version

A version for the iOS, which runs on the iPhone, iPod Touch and the iPad, was released for free on the App Store on October 27, 2008. It makes use of the multi-touch interface to move on the globe, zoom or rotate the view, and allow to select the current location using the iPhone integrated Assisted GPS. Although it previously did not support any layers apart from Wikipedia and Panoramio, version 6.2 brought KML support to add additional layers. Version 7 introduced 3D modeling of several cities.

Google Earth Plus (Discontinued in 2008)

Discontinued in December 2008, Google Earth Plus was an individual-oriented paid subscription upgrade to Google Earth that provided customers with the following features, most of which are now available in the free Google Earth.

- GPS integration: read tracks and waypoints from a GPS device. A variety of third-party applications have been created which provide this functionality using the basic version of Google Earth by generating KML or KMZ files based on user-specified or user-recorded waypoints. However, Google Earth Plus provides direct support for the Magellan and Garmin product lines, which together hold a large share of the GPS market.The Linux version of the Google Earth Plus application does not include any GPS functionality.
- Higher-resolution printing.
- Customer support via email.
- Data importer: read address points from CSV files; limited to 100 points/addresses. A feature allowing path and polygon annotations, which can be exported to KML, was formerly only available to Plus users, but was made free in version 4.0.2416.
- Higher data download speeds

Google Earth Pro

Google Earth Pro is a business-oriented upgrade to Google Earth that has more features than the Plus version. It is the most feature-rich version of Google Earth available to the public, with various additional features such as a movie maker and data importer. In addition to business-friendly features, it has also been found useful for travelers with map-making tools. Up until late January 2015, it was available for $399/year, however Google decided to make it free to the public. It is now for free and Google does not mention anything about new policy changes. The Pro version includes add-on software such as:

- Movie making.
- GIS data importer.
- Advanced printing modules.
- Radius and area measurements.

Google Earth Pro is available for Windows (NT-based versions), Mac OS X 10.4 or later.

Google Earth Enterprise

Google Earth Enterprise is a version of Google Earth designed for use by organizations whose businesses could take advantage of the program's capabilities, for example by having a globe that holds company data available for anyone in that company. As of March 20, 2015 Google has retired the Google Earth Enterprise product, with support ending March 22, 2017.

Automotive Version

An automotive version of Google Earth is available in the 2010 Audi A8.

Google Earth Plug-in

The Google Earth API is a free beta service, available for any web site that is free to consumers. The Plug-in and its JavaScript API let users place a version of Google Earth into web pages. The API enables sophisticated 3D map applications to be built. At its unveiling at Google's 2008 I/O developer conference, the company showcased potential applications such as a game where the player controlled a milktruck atop a Google Earth surface.

The Google Earth API has been deprecated as of 15 December 2014 and will remain supported until the 15th of December 2015. Google Chrome aims to end support for the Netscape Plugin API (which the Google Earth API relies on) by the end of 2016.

Controversy and Criticism

The software has been criticized by a number of special interest groups, including national officials, as being an invasion of privacy and even posing a threat to national security. The typical argument is that the software provides information about military or other critical installations that could be used by terrorists.

- Former President of India APJ Abdul Kalam expressed concern over the availability of high-resolution pictures of sensitive locations in India. Google subsequently agreed to censor such sites.
- The Indian Space Research Organisation said Google Earth poses a security threat to India, and seeks dialogue with Google officials.
- The South Korean government expressed concern that the software offers images of the presidential palace and various military installations that could possibly be used by hostile neighbor North Korea.
- In 2006, one user spotted a large topographical replica in a remote region of China. The model is a small-scale (1/500) version of the Karakoram Mountain Range, which is under the control of China but claimed by India. When later confirmed as a replica of this region, spectators began entertaining military implications.

- In 2006, Google Earth began offering detailed images of classified areas in Israel. The images showed Israel Defense Forces bases, including secret Israeli Air Force facilities, Israel's Arrow missile defense system, military headquarters and Defense Ministry compound in Tel Aviv, a top-secret power station near Ashkelon, and the Negev Nuclear Research Center. Also shown was the alleged headquarters of the Mossad, Israel's foreign intelligence service, whose location is highly classified.
- Operators of the Lucas Heights nuclear reactor in Sydney, New South Wales, Australia asked Google to censor high-resolution pictures of the facility. However, they later withdrew the request.
- In July 2007, it was reported that a new Chinese Navy Jin-class nuclear ballistic missile submarine was photographed at the Xiaopingdao Submarine Base south of Dalian.
- Hamas and the al-Aqsa Martyrs' Brigades have reportedly used Google Earth to plan Qassam rocket attacks on Israel from Gaza
- The lone surviving gunman involved in the 2008 Mumbai attacks admitted to using Google Earth to familiarise himself with the locations of buildings used in the attacks.
- Michael Finton, aka Talib Islam, used Google Earth in planning his attempted September 24, 2009, bombing of the Paul Findley Federal Building and the adjacent offices of Congressman Aaron Schock in Springfield, Illinois.
- In 2009, Google superimposed old woodblock prints of maps from 18th and 19th century Japan over Japan today. These maps marked areas inhabited by the burakumin caste, who were considered "non-humans" for their "dirty" occupations, including leather tanning and butchery. Descendants of members of the burakumin caste still face discrimination today and many Japanese people feared that some would use these areas, labeled *etamura* (穢多村, *translation: "village of an abundance of defilement""*), to target current inhabitants of them. These maps are still visible on Google Earth, but with the label removed where necessary.

Countries where Google Earth is Blocked:				
Country	**By Whom**	**Reason**	**Since When**	**Source**
Iran	Google	US government export restrictions	2007	
Morocco	Maroc Telecom, the most popular service provider	Unknown	2006	
Sudan	Google	US government export restrictions	2007	

Google Earth has been blocked by Google in Iran and Sudan since 2007 due to US government export restrictions. The program has also been blocked in Morocco since 2006 by Maroc Telecom, a major service provider in the country.

Blurred out image of the Royal Stables in The Hague, Netherlands. This has since been partially lifted.

Some citizens may express concerns over aerial information depicting their properties and residences being disseminated freely. As relatively few jurisdictions actually guarantee the individual's right to privacy, as opposed to the state's right to secrecy, this is an evolving point. Perhaps aware of these critiques, for a time, Google had Area 51 (which is highly visible and easy to find) in Nevada as a default placemark when Google Earth is first installed.

As a result of pressure from the United States government, the residence of the Vice President at Number One Observatory Circle was obscured through pixelization in Google Earth and Google Maps in 2006, but this restriction has since been lifted. The usefulness of this downgrade is questionable, as high-resolution photos and aerial surveys of the property are readily available on the Internet elsewhere. Capitol Hill also used to be pixelized in this way. The Royal Stables in The Hague, Netherlands also used to be pixelized, and are still pixelized at high zoom levels.

Critics have expressed concern over the willingness of Google to cripple their dataset to cater to special interests, believing that intentionally obscuring any land goes against its stated goal of letting the user "point and zoom to any place on the planet that you want to explore".

In the United Kingdom, critics have also argued that Google Earth has led to the vandalism of private property, highlighting the graffiti of a penis being drawn on the roof of a house near Hungerford, on the roof of Yarm School at Stockton on Tees and on the playing fields of a school in Southampton as examples of this.

In Hazleton, Pennsylvania, media attention and critics focused on Google Earth once more because of the defacing of the Hazleton Area Highschool Football field. Grass was removed to create the image of a penis approximately 35 yards long and 20 yards wide.

Late 2000s versions of Google Earth require a software component running in the background that will automatically download and install updates. Several users expressed concerns that there is not an easy way to disable this updater, as it runs without the permission of the user.

In the academic realm increasing attention has been devoted to both Google Earth and its place in the development of digital globes more generally. In particular, the International Journal of Digital Earth now features many articles evaluating and comparing the development Google Earth and its differences when compared to other professional, scientific and governmental platforms.

Elsewhere, in the Humanities and Social Sciences, Google Earth's role in the expansion of "earth observing media" has been examined. Leon Gurevitch in particular has examined the role of Google Earth in shaping a shared cultural consciousness regarding climate change and humanity's capacity to treat the earth as an engineerable object. Gurevitch has described this interface between earth representation in Google Earth and a shared cultural imaginary of geo-engineering as "Google Warming".

Copyright

Every image created from Google Earth using satellite data provided by Google Earth is a copyrighted map. Any derivative from Google Earth is made from copyrighted data which, under United States Copyright Law, may not be used except under the licenses Google provides. Google allows non-commercial personal use of the images (e.g. on a personal website or blog) as long as copyrights and attributions are preserved. By contrast, images created with NASA's globe software World Wind use The Blue Marble, Landsat or USGS layer, each of which is a terrain layer in the public domain. Works created by an agency of the United States government are public domain at the moment of creation. This means that those images can be freely modified, redistributed and used for commercial purposes.

Layers

Google Earth also features many layers as a source for information on businesses and points of interest, as well as showcasing the contents of many communities, such as Wikipedia, Panoramio and YouTube.

Borders and Labels

Contains borders for countries/provinces and shows placemarks for cities and towns.

- Borders: *Marks international borders with a thick yellow line (borders with territorial disputes with thick red lines), 1st level administrative borders (generally provinces and states) with a lavender line, and 2nd level administrative borders (counties) with a cyan line. Coastlines appear as a thin yellow line. Displays names of countries, 1st level administrative areas, and islands.*

- Labels: *Displays labels for large bodies of water, such as oceans, seas, and bays, and populated places.*

3D Imagery

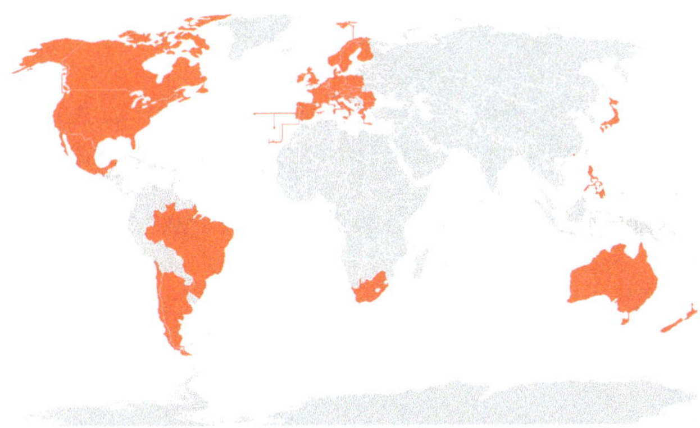

3D coverage in Google Earth

Google Earth 3D shows many 3D computer graphics building models in many cities, in these styles:

- 3D trees: *Shows many trees in Athens, Greece; Surui Forest, Brazil; Kahigaini, Kenya; Mangrove Forests, Mexico; Jedediah Smith Redwoods State Park, California*
- Photorealistic: *Shows many buildings in a realistic style, with more complex polygons and surface images.*
- Autogen: *Renders entire metropolitan areas in 3D via processing of 45 degree aerial imagery.*
- Gray: *Low-detail models of city buildings designed for computers that may not have the capability of showing the photorealistic models.*

In 2009, in a unique collaboration between Google and the Museo del Prado in Madrid, the museum selected 14 of its most important paintings to be photographed and displayed at the ultrahigh resolution of 14,000 megapixels inside the 3D version of the Prado in Google Earth and Google Maps.

In June 2012, Google announced that it will be replacing user made 3D buildings with an auto-generated 3D mesh. This will be phased in, starting with select larger cities, with the notable exception of cities such as London and Toronto which require more time to process detailed imagery of their vast number of buildings. The reason given is to have greater uniformity in 3D buildings, and to compete with other platforms already using the technology such as Nokia Here and Apple Maps.

The first 3D buildings in Google Earth were created using 3D modeling applications such as SketchUp and, beginning in 2009, Building Maker, and were uploaded to Google Earth via the 3D Warehouse.

In 2012, Google began incorporating automatically-generated 3D imagery, which displays entire areas in 3D rather than individual buildings, into the mobile and desktop versions of Google Earth, releasing coverage of 21 cities in four countries that year. By March 2015, 3D imagery covering more than 300,000 km^2 was available and by early 2016 had been expanded to hundreds of cities in over 40 countries, including every U.S. state and encompassing every continent except Antarctica.

During 2015, Hong Kong and places in the Philippines were added to the coverage.

As of February 2016, 3D imagery covering more than 495,000 km^2 was available in cities in over 40 countries, covering all continents except Antarctica.

Google Street View

Shows placemarks with 360 degree panoramic views of streets of many cities in Australia, France, the United Kingdom, Republic of Ireland, Italy, Japan, New Zealand, Spain, the United States, and recently Portugal, Brazil, the Netherlands, Taiwan, Switzerland, Canada, Mexico, Sweden, Norway, South Africa and Finland.

Weather

- Clouds – *Displays cloud cover based on data from both geostationary and low Earth-orbiting satellites. The clouds appear at their calculated elevation, determined by measuring the cloud top temperature relative to surface temperature.*
- Radar – *Displays weather radar data provided by weather.com and Weather Services International, updating every 5–6 minutes.*
- Conditions and Forecast – *Displays local temperatures and weather conditions. Clicking on an indicator displays a 2 Day Forecast (Example: Monday Morning, Monday Night, Tuesday Morning, Tuesday Night) forecast provided by weather.com.*
- Information – *Clicking Information allows users to further read up on where Google Earth gets weather information.*

Map Algebra

Map algebra is a set-based algebra for manipulating geographic data, proposed by Dr. Dana Tomlin in the early 1980s. It is a set of primitive operations in a geographic information system (GIS) which allows two or more raster layers ("maps") of similar dimensions to produce a new raster layer (map) using algebraic operations such as addition, subtraction etc.

Depending on the spatial neighborhood, GIS transformations are categorized into four classes: *local*, *focal*, *global*, and *zonal*. Local operations works on individual raster cells, or pixels. Focal operations work on cells and their neighbors, whereas global operations work on the entire layer. Finally, zonal operations work on areas of cells that share the same value. The input and output for each operator being map, the operators can be combined into a procedure or script, to perform complex tasks.

When map algebra is performed in cells from local operations, different types of operations can be used: -Arithmetic operations uses basic mathematical functions like addition, subtraction, multiplication and division. -Statistical operations uses statistical operations such as minimum, maximum, average and median. -Relational operations compares cells using functions such as greater than, smaller than or equal to. -Trigonometric operations uses sine, cosine, tangent, arcsine between two or more raster layers. -Exponential and logarithmic operations use exponent and logarithm functions.

Several major GIS systems use map algebra concepts, including ERDAS Imagine and ArcGIS. ArcGIS 10 implements Map Algebra in Python; functions are imported Python methods and Python's overloading capability is used for operators. For example, rasters can be multiplied using the "*" arithmetic operator.

Here are some examples, in MapBasic:

```
# demo for Brown's Pond data set

# Give layers

#  altitude

#  development - 0: vacant, 1: major, 2: minor, 3: houses, 4:
buildings, 5 cement

#  water - 0: dry, 2: wet, 3: pond


# calculate the slope at each location based on altitude

slope = IncrementalGradient of altitude


# identify the areas that are too steep

toosteep = LocalRating of slope
```

```
  where 1 replaces 4 5 6
  where VOID replaces ...

# create layer unifying water and development
occupied = LocalRating of development
  where water replaces VOID

notbad = LocalRating of occupied and toosteep
  where 1 replaces VOID and VOID
  where VOID replaces ... and ...

roads = LocalRating of development
  where 1 replaces 1 2
  where VOID replaces ...

nearread = FocalNeighbor of roads at 0 ... 10

aspect = IncrementalAspect of altitude

southface = LocalRating of aspect
  where 1 replaces 135 ... 225
  where VOID replaces ...

sites = LocalMinimum of nearroad and southface and notbad

sitenums = FocalInsularity of sites at 0 ... 1
```

```
sitesize = ZonalSum of 1 within sitenums

bestsites = LocalRating of sitesize
  where sitesize replaces 100 ... 300
  where VOID replaces ...
```

External Links

- osGeo-RFC-39 about Layer Algebra

Map Regression

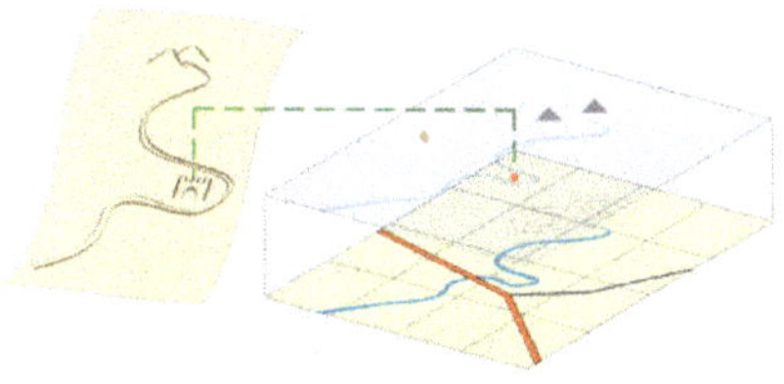

map regression - transcription

Map regression is the process of working backwards from later maps to earlier maps of the same area, to determine change or to locate past features.

Historical Map Regression

The process is mainly used in research on the history of places, sometimes termed *historic* map regression. Comparing maps of an area compiled in different periods can help reconstruct the chronology of events which have altered the natural or built environment. In archaeology, map regression can help to locate features appearing only on earlier maps and to assign building phases. It is often part of *desk based assessments* before field work is undertaken.

Methods

Map regression is performed either by comparing individual features between maps, or by re-projecting an entire map so as to fit another onto which it can then be superimposed. The process can include resolving any differences in map scale, projection, datum, or format; and the interpretation of each map in its meaning and accuracy.

Transcription

Similarities between maps, such as topography or built structures, are identified to determine where distinct features from one map are plotted on another. One manual technique is to copy features from earlier maps onto transparent overlays placed on the most recent map which provides the spatial framework. By working in sequence back in time a chronology is complied. This intensive process can promote a more considered assessment as every feature is treated individually.

Transformation

Computer based geographical information systems facilitate the scaling, rotation, and translation of an entire map to fit over another. Georeferencing establishes a congruent relationship between maps, and allows each map to be transformed and automatically re-projected as an overlay. This method is useful for the regression of maps that differ in format, for example, raster images of early maps and vector representations of modern maps.

Limitations

Features not appearing on one map but shown on another do not necessarily indicate an actual difference, due to purpose, detail, or accuracy; for example, comparing cadastral and topographic maps.

Map format can reduce confidence in a regression. 'Strip' maps, or traverse surveys, which record features in a linear path can complicate the spatial interpolation required to match a planar format.

Historical research can involve early maps of uncertain accuracy or datum. They are often small scale which leads to the magnification of any position errors in enlarging to match large scale maps. Georeferencing is not always straightforward; topography such as river courses and coastlines change over time, as do road routes and other built structures.

Historical Map Regression

The process is mainly used in research on the history of places, sometimes termed *historic* map regression. Comparing maps of an area compiled in different periods can help reconstruct the chronology of events which have altered the natural or built environment. In archaeology, map regression can help to locate features appearing only on earlier maps and to assign building phases. It is often part of *desk based assessments* before field work is undertaken.

Methods

Map regression is performed either by comparing individual features between maps, or by re-projecting an entire map so as to fit another onto which it can then be superimposed. The process can include resolving any differences in map scale, projection, datum, or format; and the interpretation of each map in its meaning and accuracy.

Transcription

Similarities between maps, such as topography or built structures, are identified to determine where distinct features from one map are plotted on another. One manual technique is to copy features from earlier maps onto transparent overlays placed on the most recent map which provides the spatial framework. By working in sequence back in time a chronology is complied. This intensive process can promote a more considered assessment as every feature is treated individually.

Transformation

Computer based geographical information systems facilitate the scaling, rotation, and translation of an entire map to fit over another. Georeferencing establishes a congruent relationship between maps, and allows each map to be transformed and automatically re-projected as an overlay. This method is useful for the regression of maps that differ in format, for example, raster images of early maps and vector representations of modern maps.

Limitations

Features not appearing on one map but shown on another do not necessarily indicate an actual difference, due to purpose, detail, or accuracy; for example, comparing cadastral and topographic maps.

Map format can reduce confidence in a regression. 'Strip' maps, or traverse surveys, which record features in a linear path can complicate the spatial interpolation required to match a planar format.

Historical research can involve early maps of uncertain accuracy or datum. They are often small scale which leads to the magnification of any position errors in enlarging to match large scale maps. Georeferencing is not always straightforward; topography such as river courses and coastlines change over time, as do road routes and other built structures.

WikiMapia

Wikimapia is a privately owned open-content collaborative mapping project, that utilizes an interactive "clickable" web map with a geographically-referenced wiki system, with the aim to mark and describe all geographical objects in the world.

Created by Alexandre Koriakine and Evgeniy Saveliev on May 2006, since then it has become a popular mapping website. The data, a crowdsourced collection of places marked by registered users and guests, has grown to over 25,000,115 objects as of August 2015, and is released under the Creative Commons License Attribution-ShareAlike (CC BY-SA).

Although the project's name is reminiscent to that of Wikipedia and that the creators share the "wiki" philosophy, it is not a part of the non-profit Wikimedia Foundation family of wikis.

Main Principles

According to the website, Wikimapia is an open-content collaborative mapping project, aimed at marking all geographical objects in the world and providing a useful description of them. It aims to, create and maintain a free, complete, multilingual and up-to-date map of the whole world. Wikimapia intends to contain detailed information about every place on Earth."

Features

Viewing

The Wikimapia website provides a Google Maps API-based interactive web map that consists of user-generated information layered on top of Google Maps satellite imagery and other resources. The navigation interface provides scroll and zoom functionality similar to that of Google Maps.

The Wikimapia layer is a collection of "objects" with a polygonal outline (buildings and lakes) and "linear features" (streets, railroads, rivers, ferry). Streets are connected by intersection points to form a street grid. Both kinds of items may have textual descriptions and photos attached to them. Viewers are able to click on any marked object or street segment to see its description. Descriptions can be searched by a built-in search tool. Tools for refining existing places according to category as well as measuring distances between objects are also available.

The interface is available in many languages, and the textual description of each item may have multiple versions in different languages.

Wikimapia maps can also be embedded on other websites.

Map Editing

The data in Wikimapia is derived from voluntary crowdsourcing. All users, registered or unregistered (*guest*), are allowed to add a place on the Wikimapia layer. Using a simple graphical editing tool, users are able to draw to draw an outline or polygon that matches the satellite image layer underneath. Each object or "tag" has specific information fields which include categories, a textual description, street address, and a related Wikipedia link. Users are likewise capable of uploading several relevant photos.

Fewer restrictions in map editing are given to registered users, who are able to edit and/or delete existing places as well as draw "*linear features*" (roads, railroads, rivers and ferry lines). A "watchlist" could be manually set up to monitor all activity or object changes made in one or more of the assigned rectangular areas on the map.

Administration

The website is maintained and developed by a small team of administrators (the *Wikimapia Team*), who introduce new features and determine further evolution course. Improvements are largely based on a feedback system from registered users through public forum discussions, bug reports and feature requests.

User Levels and Special Roles

The registered user community is largely self-organized, with users communicating through an internal message system and through a public forum. Map editing rewards the user "experience points" and milestone "awards" assigned by the system.

Registered users are automatically ranked in levels according to accumulated experience points, with higher levels gaining access to advanced tools and having less restrictions on editing activity. A registered user may be promoted to an "Advanced User" (*AU*) status as other existing *AU*s deem it fit. Additional editing and moderation tools, which includes the authority to ban users are given to an Advanced User, who is given the responsibility of countering vandalism in the map.

Special roles of maintaining the website forum, place categories, and the Wikimapia Documentation (*Docs*) are also given by the *Wikimapia Team* to some users.

Quality of Contents

The data in Wikimapia is derived from voluntary contributors who visit and add the information on the website. The textual description attached to each place object is in free format, having no restriction about style, with the exception of possessing a Neutral Point of View, where "neutral" is explained to exclude "*Feelings, opinions, experiences, words which display a personal bias or agenda, politics and/or religion*". Citing the source of the information is optional, and a link to a relevant and existing Wikipedia article is encouraged.

In spite of these recommendations, map coverage is generally uneven, with some areas, usually in developing countries, being cluttered with crude outlines, private residences, subjective evaluations or advertisements, requiring constant attention and refining by regular editors.Information can either be edited or deleted by registered users as they deem it inappropriate.

Licensing

In December 2009, Wikimapia launched an API and made its content available in several formats for non-commercial use. In December 2010, the data was announced as being available under a Non-Commercial Creative Commons license.

In May 2012, Wikimapia announced that all the content was available under Creative Commons License Attribution-ShareAlike (CC BY-SA).

Despite this, because the WikiMapia's geo-located data is largely derived from aerial imagery provided from Google Maps (whose imagery is from a number of partners including TerraMetrics, Bluesky), the dataset (and any further derivations from it) may constitute a "derived work". Whilst dependent on jurisdiction, the principle allows aerial photography companies to license their exclusive right to derive geo-data from their imagery (commercially, or under proprietary restrictions).Concerns have been raised about this, particularly from similar mapping websites.

Animal Migration Tracking

Radio-collared wolf in Yellowstone National Park

For years scientists have been tracking animals and the ways they migrate. One of the many goals of animal migration research has been to determine where the animals are going; however, researchers also want to know why they are going "there". Researchers not only look at the animals' migration but also what is between point a and point b to determine if a species is moving to new locations based on food density, a change in water temperature, and the animal's ability to adapt to these changes.

Technologies for Tracking

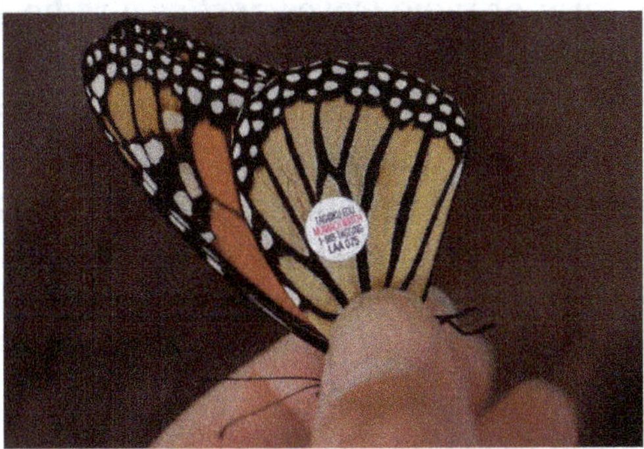

A monarch butterfly shortly after tagging at the Cape May Bird Observatory. The Observatory is one of the organisation that has a monarch identification tagging program. Plastic stickers are placed on the wing of the insect with identification information. Tracking information is used to study the migration patterns of monarchs, including how far and where they fly.

In the fall of 1803, American Naturalist John James Audubon wondered whether migrating birds returned to the same place each year. So he tied a string around the leg of a bird before it flew south. The following spring, Audubon saw the bird had indeed come back.

Scientists today still attach tags, such as metal bands, to track movement of animals. Metal bands require the re-capture of animals for the scientists to gather data; the data is thus limited to the animal's release and destination points.

Recent technologies have helped solve this problem. Some electronic tags give off repeating signals that are picked up by radio devices or satellites while other electronic tags could include archival tags (or data loggers). Scientists can track the locations and movement of the tagged animals without recapturing them using this RFID technology or satellites. These electronic tags can provide a great deal of data. However, they are more expensive than the low-tech tags that aren't electronic. Also, because of their size and weight, electronic tags may create drag on some animals, slowing them down.

Radio Tracking

The right one of these two brush-tailed rock-wallabies is wearing a radio tracking collar.

Tracking an animal by radio involves two devices. A transmitter attached to the animals sends out a signal in the form of radio waves, just as a radio station does. A scientist might place the transmitter around an animal's ankle, neck, wing, carapace, or dorsal fin. Alternatively, they may surgically implant it as internal radio transmitters have the advantage of remaining intact and functioning longer than traditional attachments, being protected from environmental variables and wear. A VHF receiver picks up the signal, just like a home radio picks up a station's signal. The receiver is usually in a truck, an ATV, or an airplane. To keep track of the signal, the scientist follows the animal using the receiver. This approach of using radio tracking can be used to track the animal manually but is also used when animals are equipped with other payloads. The receiver is used to home in on the animal to get the payload back.

Satellite Tracking

A saltwater crocodile with GPS-based satellite transmitter for migration tracking

Receivers can be placed in Earth-orbiting satellites such as ARGOS. Networks, or groups, of satellites are used to track animals. Each satellite in a network picks up electronic signals from a transmitter on an animal. Together, the signals from all satellites determine the precise location of the animal. The satellites also track the animal's path as it moves. Satellite-received transmitters fitted to animals can also provide information about the animals' physiological characteristics (e.g. temperature) and habitat use. Satellite tracking is especially useful because the scientists do not have to follow after the animal nor do they have to recover the tag to get the data on where the animal is going or has gone. Satellite networks have tracked the migration and territorial movements of caribou, sea turtles, whales, great white sharks, seals, elephants, bald eagles, ospreys and vultures. Additionally Pop-up satellite archival tags are used on marine mammals and various species of fish. There are two main systems, the above-mentioned Argos and the GPS. Thanks to these systems, conservationists can find the key sites for migratory species.

Importance

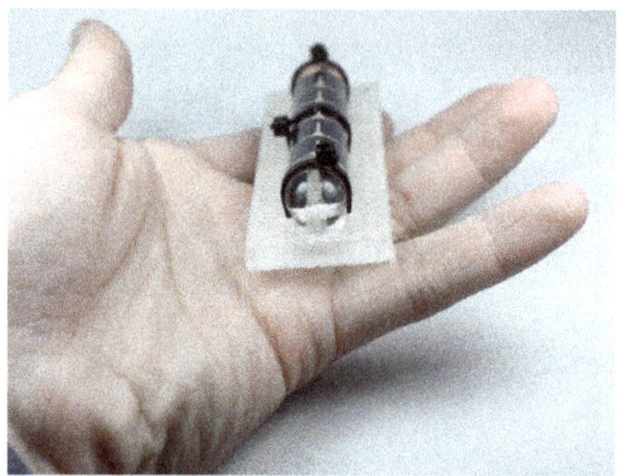

SeaTag-GEO on a turtle carrier platform for turtle tagging

Electronic tags are giving scientists a complete, accurate picture of migration patterns. For example, when scientists used radio transmitters to track one herd of caribou, they learned two important things. First, they learned that the herd moves much more than previously thought. Second, they learned that each year the herd returns to about the same place to give birth. This information would have been difficult or impossible to obtain with "low tech" tags.

Tracking migrations is an important tool to better understand and protect species. For example, Florida manatees are an endangered species, and therefore they need protection. Radio tracking showed that Florida manatees may travel as far as Rhode Island when they migrate. This information suggests that the manatees may need protection along much of the Atlantic Coast of the United States. Previously, protection efforts focused mainly in the Florida area.

In the wake of the BP oil spill, efforts in tracking animals has increased in the Gulf. Most researchers who use electronic tags have only a few options: pop-up satellite tags, archival tags, or satellite tags. Historically these tags were generally expensive and could cost several thousands of dollars per tag. However, with current advancements in technology prices are now allowing researchers to tag more animals.

Automotive Navigation System

Navigation with Gosmore, an open source routing software, on a personal navigation assistant with free map data from OpenStreetMap.

An automotive navigation system is part of the automobile controls or a third party add-on used to find direction in an automobile. It typically uses a satellite navigation device to get its position data which is then correlated to a position on a road. When directions are needed routing can be calculated. On the fly traffic information can be used to adjust the route.

Dead reckoning using distance data from sensors attached to the drivetrain, a gyroscope and an accelerometer can be used for greater reliability, as GPS signal loss and/or multipath can occur due to urban canyons or tunnels.

History

Automotive navigation systems represent a convergence of a number of diverse technologies many of which have been available for many years, but were too costly or inaccessible. Limitations such as batteries, display, and processing power had to be overcome before the product became commercially viable. Etak made an early system that used map-matching to improve on dead reckoning instrumentation. Digital map information was stored on standard cassette tapes.

- 1966: General Motors Research (GMR) was working on a non-satellite-based navigation and assistance system called DAIR (Driver Aid, Information & Routing). After initial tests GM found that it wasn't a scalable or practical way to provide navigation assistance. Decades later, however, the concept would be reborn as OnStar.
- 1980: Electronic Auto Compass with new mechanism on the Toyota Crown.
- 1981: navigation computer on the Toyota Celica. (NAVICOM)
- 1987: Toyota introduced the World's first CD-ROM-based navigation system on the Toyota Crown.
- 1990: Mazda Eunos Cosmo became the first car with built-in GPS-navigation system
- 1991: Toyota introduced GPS car navigation on the Toyota Soarer.
- 1992: Voice assisted GPS navigation system on the Toyota Celsior.
- 1995: Oldsmobile introduced the first GPS navigation system available in a United States production car, called GuideStar.
- 1995: Device called "Mobile Assistant" or short, MASS, produced by Munich-based company ComRoad AG, won the title "Best Product in Mobile Computing" on CeBit by magazine Byte. It offered turn-by-turn navigation via wireless internet connection, with both GPS and speed sensor in the car.
- 1995: Renault Safrane introduce first European CD based navigation system (CARMINAT Philips - Renault), big color display modern interface and 2D map view.
- 1995: BMW 7 series E38 second European model with the Sat Nav, using pretty same software and hardware (Philips - Renault) like Renault Safrane, build in magnetometer, more features available from the menu like: AUX ventilation, TV or Phone
- 1997: Navigation system using Differential GPS developed as a factory-installed option on the Toyota Prius

- 1998: First DVD-based navigation system introduced on the Toyota Progres
- 2000: The United States made a more accurate GPS signal available for civilian use.
- 2003: Toyota introduced the first Hard disk drive-based navigation system and the industry's first DVD-based navigation system with a built-in Electronic throttle control
- 2007: Toyota introduced Map on Demand, a technology for distributing map updates to car navigation systems, developed as the first of its kind in the world
- 2008: World's first navigation system-linked brake assist function and Navigation system linked to Adaptive Variable Suspension System (NAVI/AI-AVS) on Toyota Crown

Technology

The road database is a vector map. Street names or numbers and house numbers are encoded as geographic coordinates so that the user can find some desired destination by street address.

Points of interest (waypoints) are stored with their geographic coordinates. Formats are almost uniformly proprietary; there is no industry standard for satellite navigation maps, although some companies are trying to address this with SDAL and NDS.

Map data vendors such as Tele Atlas and Navteq create the base map in a Geographic Data Files format, but each electronics manufacturer compiles it in an optimized, usually proprietary format. GDF is not a CD standard for car navigation systems. GDF is used and converted onto the CD-ROM in the internal format of the navigation system.

CARiN Database Format (CDF) is a proprietary navigation map format created by Philips Car Systems (this branch was sold to Mannesman VDO, which became VDO/Dayton in 1998, then Siemens VDO in 2002, then Continental in 2007) and is used in a number of navigation-equipped vehicles. The 'CARiN' portmanteau is derived from Car Information and Navigation.

SDAL is a proprietary map format published by Navteq, who released it royalty free in the hope that it would become an industry standard for digital navigation maps. Vendors who used this format include:

- Microsoft
- Magellan
- Pioneer
- Panasonic

- Clarion
- InfoGation

The format has not been very widely adopted by the industry.

Navigation Data Standard (NDS)

The Navigation Data Standard (NDS) initiative, is an industry grouping of car manufacturers, navigation system suppliers and map data suppliers whose objective is the standardization of the data format used in car navigation systems, as well as allow a map update capability. The NDS effort began in 2004 and became a registered association in 2009. Standardization would improve interoperability, specifically by allowing the same navigation maps to be used in navigation systems from 20 manufacturers. Companies involved include BMW, Volkswagen, Daimler, Renault, ADIT, Aisin AW, Alpine Electronics, Navigon, Navis-AMS, Bosch, DENSO, Mitsubishi, Harman International Industries, Panasonic, PTV, Continental AG, Clarion, Navteq, Navinfo, TomTom and Zenrin.

Media

The road database may be stored in solid state read-only memory (ROM), optical media (CD or DVD), solid state flash memory, magnetic media (hard disk), or a combination. A common scheme is to have a base map permanently stored in ROM that can be augmented with detailed information for a region the user is interested in. A ROM is always programmed at the factory; the other media may be preprogrammed, downloaded from a CD or DVD via a computer or wireless connection (bluetooth, Wi-Fi), or directly used utilizing a card reader.

Some navigation device makers provide free map updates for their customers. These updates are often obtained from the vendor's website, which is accessed by connecting the navigation device to a PC.

Real-Time Data

Some systems can receive and display information on traffic congestion using either TMC, RDS, or by GPRS/3G data transmission via mobile phones.

Integration and other Functions

- The color LCD screens on some automotive navigation systems can also be used to display television broadcasts or DVD movies.
- A few systems integrate (or communicate) with mobile phones for hands-free talking and SMS messaging (i.e., using Bluetooth or Wi-Fi).

- Automotive navigation systems can include personal information management for meetings, which can be combined with a traffic and public transport information system.

Original Factory Equipment

Many vehicle manufacturers offer a GPS navigation device as an option in their vehicles. Customers whose vehicles did not ship with GPS can therefore purchase and retrofit the original factory-supplied GPS unit. In some cases this can be a straightforward "plug-and-play" installation if the required wiring harness is already present in the vehicle. However, with some manufacturers, new wiring is required, making the installation more complex.

The primary benefit of this approach an integrated and factory-standard installation. Many original systems also contain a gyrocompass or accelerometer and may accept input from the vehicle's speed sensors, thereby allowing them to navigate via dead reckoning when a GPS signal is temporarily unavailable. However, the costs can be considerably higher than other options.

SMS

Establishing points of interest in real-time and transmitting them via GSM cellular telephone networks using the Short Message Service (SMS) is referred to as Gps2sms. Some vehicles and vessels are equipped with hardware that is able to automatically send an SMS text message when a particular event happens, such as theft, anchor drift or breakdown. The receiving party (e.g., a tow truck) can store the waypoint in a computer system, draw a map indicating the location, or see it in an automotive navigation system.

Aerial Video

USGS aerial photography crew gathers data following the landfall of a hurricane.

Digital video and recording system used aboard aircraft during coastal oblique video and photography missions.

Aerial video is an emerging form of data acquisition for scene understanding and object tracking. The video is captured by low flying aerial platforms that integrate Global Positioning Systems (GPS) and automated image processing to improve the accuracy and cost-effectiveness of data collection and reduction. Recorders can incorporate in-flight voice records from the cockpit intercom system. The addition of audio narration is an extremely valuable tool for documentation and communication. GPS data is incorporated with a text-captioning device on each video frame. Helicopter platforms enable "low and slow" flights, acquiring a continuous visual record without motion blur.

Innovations in remote sensing cameras have allowed the identification of objects that could not have been previously identified. Pipeline and power corridors and their infrastructure can be documented with digital media recording. Video Mapping System is an example of how this technology is used today.

Since the 1980s, aerial videography has seen increased use in applications where its advantages over traditional photography (lower cost and immediate availability of data) outweigh its disadvantages (poorer spatial resolution and difficulty of analysis due to lack of stereo imaging) (Mausel et al. 1992; Meisner 1986). King (1995) provides a comprehensive review of the evolution of video sensors and their applications, many of which focused on:

1. The measurement of transient phenomena such as wildlife populations (Sidle and Ziewits 1990; Strong and Cowardin 1995) and pest infestations (Everitt et al. 1994);

2. Mapping of dynamic land features such as wetland plant communities (Jennings et al. 1992) and coastallandforms (Eleveld et al. 2000);

3. Land cover mapping in remote areas with limited existing aerial photography and poor infrastructure(Marshet al. 1994; Slaymaker and Hannah 1997).

References

- Brantingham, Paul J.; Brantingham, Patricia L., eds. (1981). Environmental Criminology. Waveland Press. ISBN 0-88133-539-8.
- Kelling, George; Coles, Catherine (1997) [1996]. Fixing Broken Windows: Restoring Order and Reducing Crime in Our Communities. New York: Simon & Schuster. ISBN 0-684-83738-2.
- Longley; et al. Geographic Information Systems and Science. John Wiley & Sons, Inc. pp. 414–7. ISBN 978-0-470-72144-5.
- "Google Earth adds new 3D imagery in 21 cities to its 11,000 guided tours of our planet". Retrieved 24 July 2016.
- "Autoradio GPS Android pas cher, Caméra radar de recul - Player Top". www.player-top.fr. Retrieved 2016-07-18.
- "Wikimapia Docs; Help/FAQ". Wikimapia.org. Retrieved 24 August 2015. Is it WikiMapia or Wikimapia? This website is usually more referred to as Wikimapia.
- Jason D. O'Grady (2 February 2009). "Google Earth 5 released for the Mac". ZDNet. Retrieved January 7, 2015.
- Panzarino, Matthew (31 October 2012). "Google Earth 7 Gets 11 3D Cities and 11,000 Virtual Tours". The Next Web. Retrieved 2015-01-07.
- Rose, Robert (2015-02-11). "Good News Everyone! Google Earth Pro Now Free!". Mapshole. Retrieved 22 February 2015.
- "NDS Partners, NDS Association". NDS Association. Retrieved 2015-02-13. External link in |publisher= (help)
- "Google Lat Long: See the seafloor like never before on World Oceans Day". Google-latlong.blogspot.com. 2011-06-08. Retrieved 2013-06-15.
- "Keyhole Markup Language — Google Developers". Developers.google.com. 2012-03-01. Retrieved 2013-06-15.
- Thornton, James (2013-06-05). "Google SketchUp – Download". Google-sketchup.en.softonic.com. Retrieved 2013-06-15.
- grayaudio on Mar 15, 2010. "World's Highest-Resolution Satellite Imagery". HotHardware. Retrieved 2013-06-09.
- Understanding the Archaeology of Landscapes (PDF), English Heritage, 2007, Part 4, p.30, retrieved 28 December 2013

6

Interdisciplinary Aspects of Geographic Information Systems

A geographic information system is an interdisciplinary subject. This section will provide a glimpse of the related fields of geographic information system. GIS and public health, GIS and aquatic science, GIS and hydrology and GIS and environmental governance are some of the aspects elucidated in the text.

GIS and Public Health

Geographic information systems (GISs) and geographic information science (GIScience) combine computer-mapping capabilities with additional database management and data analysis tools. Commercial GIS systems are very powerful and have touched many applications and industries, including environmental science, urban planning, agricultural applications, and others.

Public health is another focus area that has made increasing use of GIS techniques. A strict definition of public health is difficult to pin down, as it is used in different ways by different groups. In general, public health differs from personal health in that it is (1) focused on the health of populations rather than of individuals, (2) focused more on prevention than on treatment, and (3) operates in a mainly governmental (rather than private) context. These efforts fall naturally within the domain of problems requiring use of spatial analysis as part of the solution, and GIS and other spatial analysis tools are therefore recognized as providing potentially transformational capabilities for public health efforts.

This article presents some history of use of geographic information and geographic information systems in public health application areas, provides some examples showing the utilization of GIS techniques in solving specific public health problems, and finally addresses several potential issues arising from increased use of these GIS techniques in the public health arena.

History

Public health efforts have been based on analysis and use of spatial data for many years. Dr. John Snow (physician), often credited as the father of epidemiology, is arguably the most famous of those examples. Dr. Snow used a hand-drawn map to analyze the geographic locations of deaths related to cholera in London in the mid-1850s. His map, which superimposed the locations of cholera deaths with those of public water supplies, pinpointed the Broad Street pump as the most likely source of the cholera outbreak. Removal of the pump handle led to a rapid decline in the incidence of cholera, helping the medical community to eventually conclude that cholera was a water-borne disease.

Dr. Snow's map showing cholera cases in London during the epidemic of 1854.

Dr. Snow's work provides an indication of how a GIS could benefit public health investigations and other research. He continued to analyze his data, eventually showing that the incidence rate of cholera was also related to local elevation as well as soil type and alkalinity. Low-lying areas, particularly those with poorly draining soil, were found to have higher incidence rates for cholera, which Dr. Snow attributed to the pools of water that tended to collect there, again showing evidence that cholera was in fact a water-borne disease (rather than one borne by 'miasma' as was commonly believed at the time.

This is an early example of what has come to be known as disease diffusion mapping, an area of study based on the idea that a disease starts from some source or central point and then spreads throughout the local area according to patterns and conditions there. This is another area of research where the capabilities of a GIS have been shown to be of help to practitioners.

GIS for Public Health

Today's public health problems are much larger in scope than those Dr. Snow faced, and researchers today depend on modern GIS and other computer mapping applications to assist in their analyses. For example, see the map to the right depicting death rates from heart disease among white males above age 35 in the US between 2000 and 2004.

Public health informatics (PHI) is an emerging specialty which focuses on the application of information science and technology to public health practice and research. As part of that effort, a GIS – or more generally a spatial decision support system (SDSS) – offers improved geographic visualization techniques, leading to faster, better, and more robust understanding and decision-making capabilities in the public health arena.

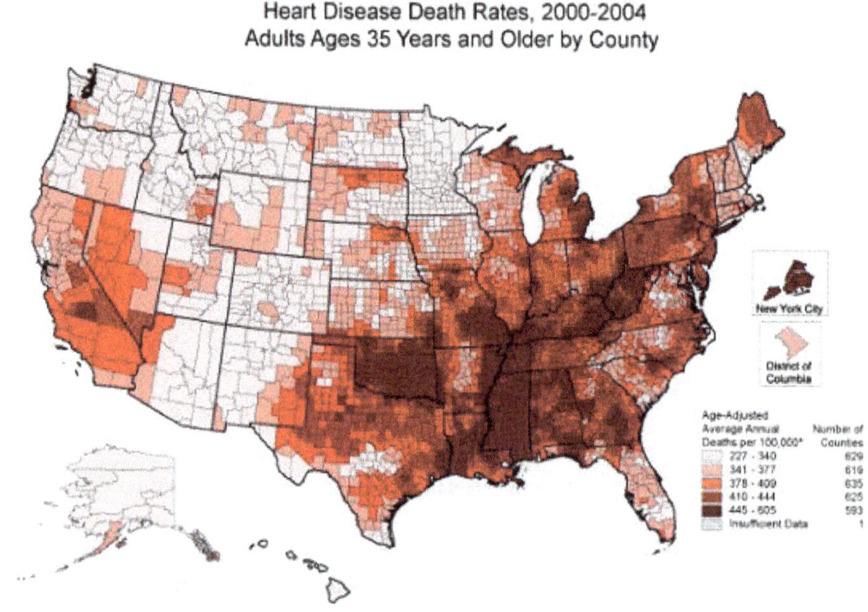

More modern disease map showing deaths from heart disease among white males in the US from 2000–2004.

For example, GIS displays have been used to show a clear relationship between clusters of emergent Hepatitis C cases and those of known intravenous drug users in Connecticut. Causality is difficult to prove conclusively – collocation does not establish causation – but confirmation of previously established causal relationships (like intravenous drug use and Hepatitis C) can strengthen acceptance of those relationships, as well as help to demonstrate the utility and reliability of GIS-related solution techniques. Conversely, showing the coincidence of potential causal factors with the ultimate effect can help suggest a potential causal relationship, thereby driving further investigation and analysis (source needed?).

Alternately, GIS techniques have been used to show a lack of correlation between causes and effects or between different effects. For example, the distributions of both birth defects and infant mortality in Iowa were studied, and the researchers found no relationship in those data. This led to the conclusion that birth defects and infant mortality are likely unrelated, and are likely due to different causes and risk factors.

GIS can support public health in different ways as well. First and foremost, GIS displays can help inform proper understanding and drive better decisions. For example, elimination of health disparities is one of two primary goals of Healthy People 2010, one of the preeminent public health programs in existence today in the US. GIS can play a significant role in that effort, helping public health practitioners identify areas of disparities or inequities, and ideally helping them identify and develop solutions to address those shortcomings. GIS can also help researchers integrate disparate data from a wide variety of sources, and can even be used to enforce quality control measures on those data. Much public health data is still manually generated, and is therefore subject to human-generated mistakes and miscoding. For example, geographic analysis of health care data from North Carolina showed that just over 40% of the records contained errors of some sort in the geographic information (city, county, or zip code), errors that would have gone undetected without the visual displays provided by GIS. Correction of these errors led not only to more correct GIS displays, but also improved ALL analyses using those data.

Issues with GIS for Public Health

There are also concerns or issues with use of GIS tools for public health efforts. Chief among those is a concern for privacy and confidentiality of individuals. Public health is concerned about the health of the population as a whole, but must use data on the health of individuals to make many of those assessments, and protecting the privacy and confidentiality of those individuals is of paramount importance. Use of GIS displays and related databases raises the potential of compromising those privacy standards, so some precautions are necessary to avoid pinpointing individuals based on spatial data. For example, data may need to be aggregated to cover larger areas such as a zip code or county, helping to mask individual identities. Maps can also be constructed at smaller scales so that less detail is revealed. Alternately, key identifying features (such as the road and street network) can be left off the maps to mask exact location, or it may even be advisable to intentionally offset the location markers by some random amount if deemed necessary.

It is well established in the literature that statistical inference based on aggregated data can lead researchers to erroneous conclusions, suggesting relationships that in fact do not exist or obscuring relationships that do in fact exist. This issue is known as the modifiable areal unit problem. For example, New York public health officials worried that cancer clusters and causes would be misidentified after they were forced to post maps showing cancer cases by ZIP code on the internet. Their assertion was that ZIP codes were designed for a purpose unrelated to public health issues, and so use of these arbitrary boundaries might lead to inappropriate groupings and then to incorrect conclusions.

Summary

Use of GIS in public health is an application area still in its infancy. Like most new applications, there is a lot of promise, but also a lot of pitfalls that must be avoided along the way. Many researchers and practitioners are concentrating of this effort, hoping that the benefits outweigh the risks and the costs associated with this emerging application area for modern GIS techniques.

GIS and Aquatic Science

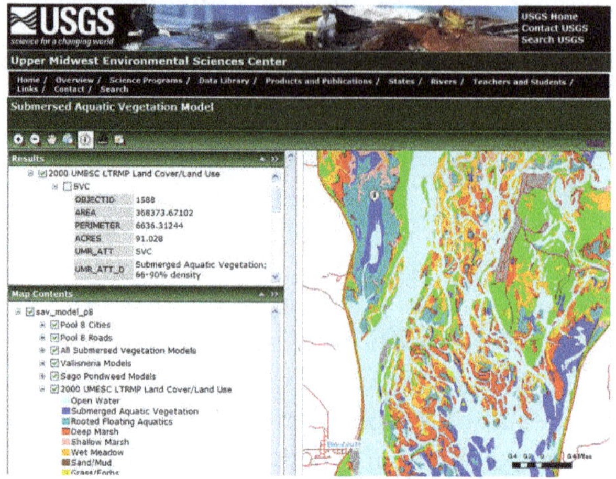

ArcGIS Server website depicting submersed aquatic vegetation.

Geographic Information Systems (GIS) has become an integral part of aquatic science and limnology. Water by its very nature is dynamic. Features associated with water are thus ever-changing. To be able to keep up with these changes, technological advancements have given scientists methods to enhance all aspects of scientific investigation, from satellite tracking of wildlife to computer mapping of habitats. Agencies like the US Geological Survey, US Fish and Wildlife Service as well as other federal and state agencies are utilizing GIS to aid in their conservation efforts.

GIS is being used in multiple fields of aquatic science from limnology, hydrology, aquatic botany, stream ecology, oceanography and marine biology. Applications include using satellite imagery to identify, monitor and mitigate habitat loss. Imagery can also show the condition of inaccessible areas. Scientists can track movements and develop a strategy to locate locations of concern. GIS can be used to track invasive species, endangered species, and population changes.

One of the advantages of the system is the availability for the information to be shared and updated at any time through the use of web-based data collection.

GIS and Fish

In the past, GIS was not a practical source of analysis due to the difficulty in obtaining spatial data on habitats or organisms in underwater environments. With the advancement of radio telemetry, hydroacoustic telemetry and side-scan sonar biologists have been able to track fish species and create databases that can be incorporated into a GIS program to create a geographical representation. Using radio and hydro-acoustic telemetry, biologists are able to locate fish and acquire relatable data for those sites, this data may include substrate samples, temperature, and conductivity. Side-scan sonar allows biologists to map out a river bottom to gain a representation of possible habitats that are used. These two sets of data can be overlaid to delineate the distribution of fish and their habitats for fish. This method has been used in the study of the pallid sturgeon.

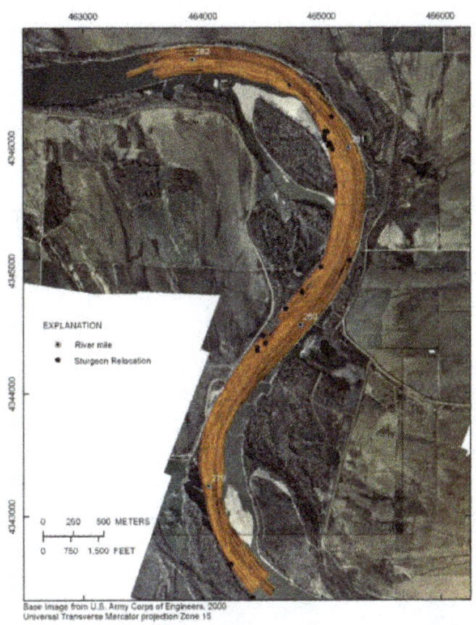

USGS sidescan radar image over base image from Army Corps of Engineers, indicating sturgeon location and river mile.

Over a period of time large amounts of data are collected and can be used to track patterns of migration, spawning locations and preferred habitat. Before, this data would be mapped and overlaid manually. Now this data can be entered into a GIS program and be layered, organized and analyzed in a way that was not possible to do in the past. Layering within a GIS program allows for the scientist to look at multiple species at once to find possible watersheds that are shared by these species, or to specifically choose one species for further examination. The US Geological Survey (USGS) in, cooperation with other agencies, were able to use GIS in helping map out habitat areas and movement patterns of pallid sturgeon. At the Columbia Environmental Research Center their effort relies on a customized ArcPad and ArcGIS, both ESRI (Environmental Systems Research Institute) applications, to record sturgeon movements to streamline data collection. A relational database was developed to manage tabular data for each individual sturgeon, including initial capture and reproductive physiology. Movement maps can be created for individual sturgeon. These maps help track the movements of each sturgeon through space and time. This allowed these researchers to prioritize and schedule field personnel efforts to track, map, and re-capture sturgeon.

GIS and Macrophytes

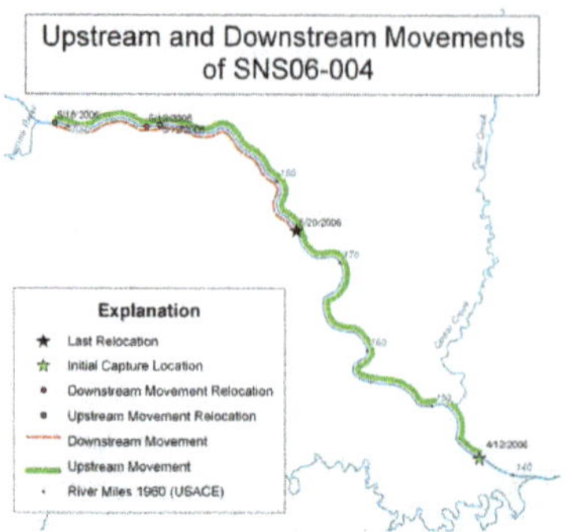

Map created from GIS database depicting the movements of individual sturgeon.

Surveyed (left) and predicted (right) distributions of submersed aquatic vegetation distribution Upper Mississippi River in 1989. The survey data were from the land cover/land use geographic information created by the U.S. Geological Survey Upper Midwest Environmental Sciences Center on the basis of interpretation of aerial photography of 1989.

Macrophytes are an important part of healthy ecosystems. They provide habitat, refuge, and food for fish, wildlife, and other organisms. Though natural occurring species are of great interest so are the invasive species that occur alongside these in our environment. GIS is being used by agencies and their respective resource managers as a tool to model these important macrophyte species. Through the use of GIS resource managers can assess the distributions of this important aspect of aquatic environments through a spatial and temporal scale. The ability to track vegetation change through time and space to make predictions about vegetation change are some of the many possibilities of GIS. Accurate maps of the aquatic plant distribution within an aquatic ecosystem are an essential part resource management.

It is possible to predict the possible occurrences of aquatic vegetation. For example, the USGS has created a model for the American wild celery (Vallisneria americana) by developing a statistical model that calculates the probability of submersed aquatic vegetation. They established a web link to an Environmental Systems Research Institute (ESRI) ArcGIS Server website *Submersed Aquatic Vegetation Model to make their model predictions available online. These predictions for distribution of submerged aquatic vegetation can potentially have an effect on foraging birds by creating avoidance zones by humans. If it is known where these areas are, birds can be left alone to feed undisturbed. When there are years where the aquatic vegetation is predicted to be limited in these important wildlife habitats, managers can be alerted.

Invasive species have become a great conservation concern for resource managers. GIS allows managers to map out plant locations and abundances. These maps can then be used to determine the threat of these invasive plants and help the managers decide on management strategies. Surveys of these species can be conducted and then downloaded into a GIS system. Coupled with this, native species can be included to determine how these communities respond with each other. By using known data of preexisting invasive species GIS models could predict future outbreaks by comparing biological factors. The Connecticut Agricultural Experiment Station Invasive Aquatic Species Program (CAES IAPP) is using GIS to evaluate risk factors. GIS allows managers to georeference plant locations and abundance. This allows for managers to display invasive communities alongside native species for study and management.

GIS and Hydrology

Geographic information systems (GISs) have become a useful and important tool in hydrology and to hydrologists in the scientific study and management of water resources. Climate change and greater demands on water resources require a more knowledgeable disposition of arguably one of our most vital resources. As every hydrologist knows, water is constantly in motion. Because water in its occurrence varies spatially and temporally throughout the hydrologic cycle, its study using GIS is especially practical. GIS systems previously were mostly static in their geospatial representation of hydrologic features. Today, GIS platforms have become increasingly dynamic, narrowing the gap between historical data and current hydrologic reality.

The elementary water cycle has inputs equal to outputs plus or minus change in storage. Hydrologists make use of a hydrologic budget when they study a watershed. A watershed is a spatial area, and the occurrence of water throughout its space varies by time. In the hydrologic budget are inputs such as precipitation, surface flows in, and groundwater flows in. Outputs are evapotranspiration, infiltration, surface runoff, and surface/groundwater flows out. All of these quantities, including storage, can be measured or estimated, and their characteristics can be graphically displayed in GIS and studied.

As a subset of hydrology, hydrogeology is concerned with the occurrence, distribution, and movement of groundwater. Moreover, hydrogeology is concerned with the manner in which groundwater is stored and its availability for use. The characteristics of groundwater can readily be input into GIS for further study and management of water resources. Because 98% of the world's available freshwater is groundwater, the need to keep a closer eye on its disposition is readily apparent.

GIS in Surface Water

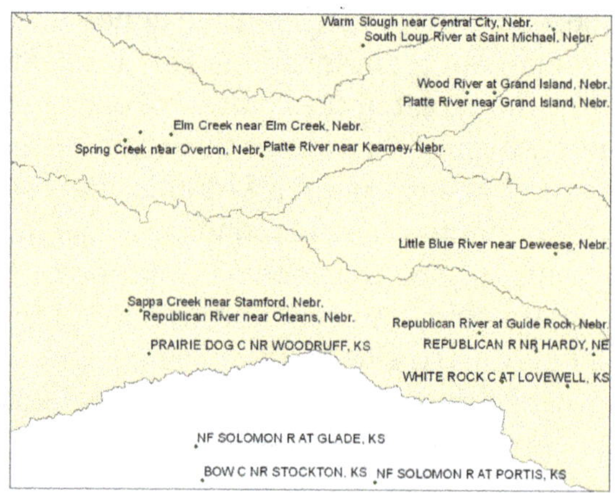

USGS Real-Time streamflow gage locations with hyperlinks within a GIS to the data

It is possible to access historical and real time streamflow data via the Internet. Embedded within a GIS are layers with stream locations and gage or measuring/monitoring sites. It's also possible to link radio transmitted and remotely sensed (Remote Sensing) data in GIS. Historical and real time data are available from the United States Geological Survey (USGS) in the form of gage height and streamflow or discharge in cubic feet per second. Within a GIS, it's possible to direct link via the Internet to real time data. Other sources of data for flood information and water quality come from the National Weather Service (NWS) and United States Environmental Protection Agency (EPA). All these data are available for analysis within GIS, providing a spatial representation of what would otherwise be data in a table type format.

GIS is much more capable of displaying data spatially than temporally. Within one GIS, ESRI's ArcGIS for example, is it possible to delineate a watershed. Digital elevation model (DEM) data are layered with hydrographic data so that the boundaries of a watershed may be determined. Watershed delineation aids the hydrologist or water resource manager in understanding where runoff from precipitation or snowmelt will eventually drain. In the case of snowmelt, snowpack coverage may be determined from ground stations or remotely sensed observers and input into GIS to determine or predict how much water can be counted on to be available for use by cities, agriculture, and environmental habitat.

Another useful application for GIS regards precipitation, but other hydrologic data (evapotranspiration, infiltration, and groundwater) may be treated similarly. Precipitation is an area event measured using data from point locations. The difficulty in using point data lies in extrapolating these point measurements to areas. One useful method to extrapolate data is to construct Thiessen polygons which assess the distance and geometry of points in a plane and determines representative areas for which to assign precipitation values. GIS applications like ArcGIS are capable of constructing Thiessen polygons, and other methods of determining area precipitation are viable with GIS as well.

A step up in complexity from manual analysis of select spatially depicted hydrologic data is to display a representative version of hydrologic reality and perhaps merge it with a numerical or other model which might predict what might happen say x amount of rainfall occurs or to forecast, for example, runoff following the passage of an approaching weather system. One such method to do this would be to connect a GIS data model with a simulation model. The GIS data model has all the relevant surface water features with attributes that describe historical or current hydrologic data. The data model structures all the pertinent data to arrive at a representative depiction of hydrologic reality for display and analysis. One data model which does this is Arc Hydro, created cooperatively by ESRI and the Center for Research in Water Resources (CRWR) at the University of Texas at Austin to work within ESRI's ArcGIS. It is important to understand the data model does not predict as this is the function of the simulation model that Arc Hydro might feed. The simulation model is very complex and beyond the scope of this article.

By synthesizing GIS technology with hydrologic data, it has become possible to elucidate the effects of watershed-scale land-use and land-cover changes. For example, with growing pressures on water resources there is a strong interest in how forestation affects water yields. GIS and remote sensing facilitate quantifying long-term changes in forest cover since aerial photography records are available across much of the United States since as early as the 1930s. Even earlier than the 1930s the USGS started systematically gauging many watersheds throughout the country. Once long-term land-cover trends have been quantified in a gauged watershed, it becomes possible to statistically compare the long-term land-cover changes with the land-use changes to determine, for example, if forestation is actually reducing streamflow as is widely perceived. Thus using GIS data together with hydrology data can allow for knowledge based water resources decision making at far lower costs than traditional methods!

GIS in Groundwater

As mentioned earlier, 98% of the available freshwater (negating polar and glacial ice) for human and environmental uses is in groundwater. In the United States, about ¼ of the water used for personal, commercial/industrial, and irrigation uses comes from groundwater. With increasing demands placed on surface water resources, it is likely the demand for groundwater will increase. In some places, this resource has already been severely tapped, and even mismanaged. An example here is the surface water decline in the Republican River watershed of Nebraska and Kansas where over-pumping of groundwater for irrigation in Nebraska has depleted surface water available for downstream flow and use in Kansas resulting in a lawsuit by that state against the state of Nebraska. Although not as apparent as surface water flow, groundwater can also be characterized spatially in a GIS and analyzed by scientists and natural resource managers.

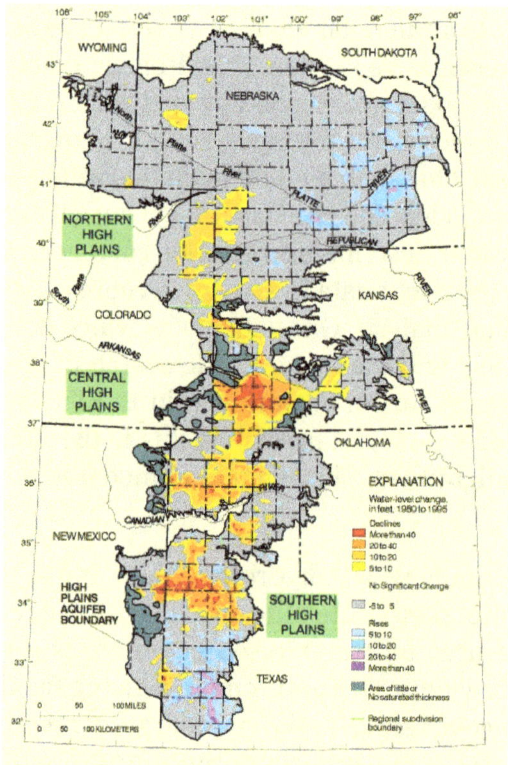

Groundwater level change of the High Plains Aquifer, 1980–95

It can be argued that the depiction of groundwater is an even more complex task than that of surface water. The two resources are by no means disjoint, as knowing where surface water recharges groundwater and where groundwater flows supply surface water is an important aspect of the hydrologic cycle. Hydrogeology is especially well suited to GIS. Groundwater moves much more slowly than surface water, on the order of less than a meter per day up to perhaps a hundred meters per day, and is 3-dimensional in flow. In contrast, surface water flows much faster and is more two-dimensional. Groundwater flow is a function of geology and "head," the total potential energy at a location. Groundwater flows from higher head to lower head at a travel rate and flow path dictated by geology. Head values, geology, groundwater flow direction, even water table height and location of aquifers are among the quantities which may be presented spatially in GIS and used for analysis, management of water availability and water quality, and land use practices.

A very large amount of data from wells is available such as location, depth to water, stratigraphy, water quality and chemistry, aquifer characteristics, and the list goes on. The volume of data can be managed in a GIS and manipulated to display spatial characteristics for analysis and water resource planning. For example, in a simple application of GIS, the effect of a new well can be studied on the existing groundwater and surface water. The results of such a study can be used by decision makers to determine whether or not to proceed with drilling.

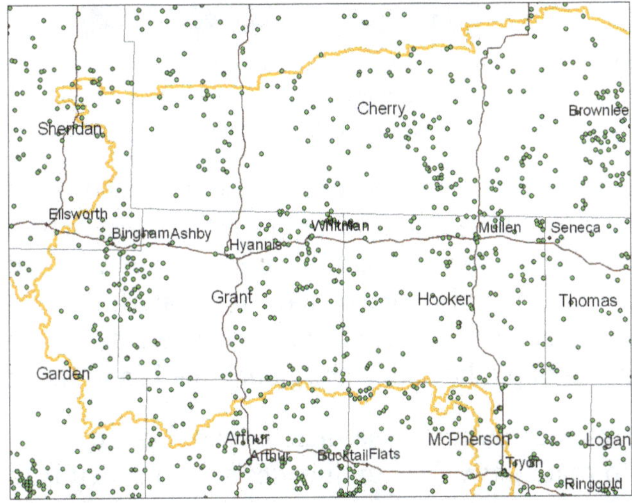

Nebraska Sandhills: registered well locations in the Upper Loup basin

An especially useful application of GIS concerns water quality in groundwater. For construction/situating of industrial plants, landfills, agricultural activities, and other potential groundwater contamination sources, it is useful to know how existing groundwater supplies could be affected or would be at risk of impact. Further, in the case of groundwater contamination and the need for subsequent containment and cleanup of the contaminant, an existing framework of the groundwater system would be valuable in planning remediation measures. This GIS could be the front end to a groundwater modeling simulation devised to fully capture the contaminant. An additional example concerning the use of GIS addresses a common problem associated with groundwater pumping and land subsidence or intrusion in coastal areas. Areas that have been overpumped of groundwater can subside, and when near the sea, this may invite flooding. Also, overpumping of groundwater in coastal regions may bring a different problem, such as the case in California where salt-water intrusion has compromised the aquifer. Generally, a salt water interface inland of the coast extends below the land surface dependent on the distance from the coast. Overpumping can bring the salt water interface to a higher position and contaminate an aquifer. A careful study and management of groundwater within GIS or with modeled GIS data can forestall or alleviate these problems.

GIS and Environmental Governance

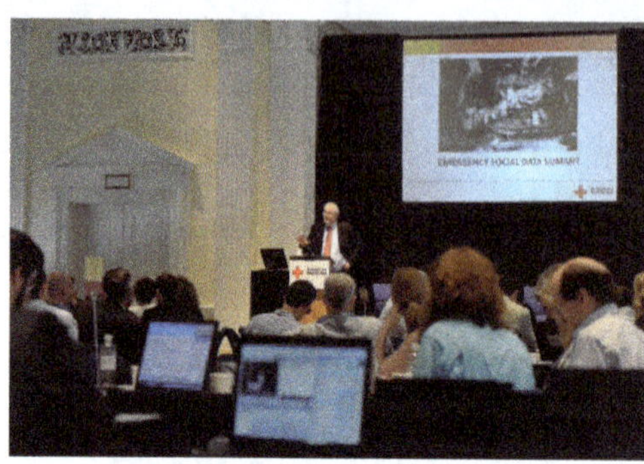

U.S. Federal Emergency Management Agency (FEMA) Administrator W. Craig Fugate speaking at a Red Cross seminar on using social media during natural disasters. GIS has an integral role to play in such agendas.

Geographic Information Systems (GIS) is a commonly used tool for environmental management, modelling and planning. As simply defined by Michael Goodchild, GIS is as 'a computer system for handling geographic information in a digital form'. In recent years it has played an integral role in participatory, collaborative and open data philosophies. Social and technological evolutions have elevated 'digital' and 'environmental' agendas to the forefront of public policy, the global media and the private sector.

Government departments routinely use digital spatial platforms to plan and model proposed changes to road networks, building design, greenbelt land, utility provision, crime prevention, energy production, waste management and security. Non-profit organizations also incorporate geospatial and web-mapping approaches into political campaigns to lobby governments, to protest against socially or environmentally harmful companies, and to generate public support. Private business, whether in land management, resource extraction, retail, manufacturing or social media for example, also incorporate GIS into overall profit-making strategies.

Citizen Science and GIS

Citizen science is part of the wider emphasis upon public involvement in expert fields across Western democracies. The term is 'often used to describe communities or networks of citizens who act as observers in some domain of science'. Although more narrowly used to describe the shift to specifically user-generated forms of knowledge creation, it has been routinely invoked in both the public participatory GIS and environmental governance literature at large.

The Secretary of the US Navy, Ray Mabus is briefed on the Deepwater Horizon oil spill response. A web-based GIS is visible in the background. The NOAA-developed Environmental Response Management Application (ERMA) was designed to assist resource managers post-spill.

National Audubon Society: Gulf Spill Bird Tracker

Much of the citizen science literature is grounded in wildlife study. For example, Goodchild references the National Audubon Society's Christmas Bird Count (CBC) as a classic case of citizen science in action. Each year over the winter period, the American conservation organization encourages volunteer bird-watchers to gather information on the number of bird species in their local area. Once field data has been collected, each volunteer is able to submit their bird sightings into an online database, for the benefit of both scientific researchers and bird enthusiasts. The eBird project – enabling the general public to explore a range of map- and chart-based bird species datasets - is a result of these yearly mass volunteer events.

Of particular interest is the Gulf Spill Bird Tracker; an interactive sightings map for ten species deemed at risk from the Deepwater Horizon oil spill in 2010. Gulf Coast bird watchers were encouraged (at the time when it was live) to submit their sightings of a range of at-danger birds (such as the brown pelican, roseate spoonbill and the Wilson's plover), to help aid the clean-up operation, and pin-point beaches most affected by the oil spill. The National Audubon Society has been deeply involved in the Gulf Oil Response since the disaster, and has a dedicated program to co-ordinate resources, liaise with local government, and deploy equipment post-spill. Their 6-month report brought together some of these key factors. Not only was the National Audubon Society's citizen science initiative highlighted as the 'backbone...for understanding the [immediate] impact of the disaster', but also for long-term efforts to monitor the health of imperilled species in the Gulf Coast region. Moreover, their grassroots ethos has mobilized a vast number of Gulf volunteers to 'urge elected officials and government agencies to hold polluters like BP accountable', for the financial, environmental, economic and social costs associated with such disasters. This is perhaps the most obvious example of web-based mapping software (a more 'citizen-friendly' form of GIS) and environmental governance discourses colliding head on. The notion of volunteered, user-generated, citizen data is the guiding mantra for such projects, and the cornerstone of any wider attempts to lobby national governments, engage with local community groups, and generate scientific research.

Mapping for Change

Another example of citizen science and GIS in action is taken from inside the academy. University College London (UCL) and London 21 sustainability network's Mapping for Change initiative has encouraged voluntary groups, local authorities and development agencies to build map-based projects to support political, social and environmental aims. They even provide a noise mapping toolkit on the Mapping for Change website itself, designed to help local communities gather evidence of intrusive and unwanted environmental nuisances and hazards. The Royal Docks community in London has used such a toolkit to help present their concerns to the Greater London Authority Environment Committee over plans to expand London City Airport. Armed with sound meters, survey sheets and access to an online mapping platform, residents were able to monitor noise levels; from overhead planes and passing motor vehicles, to birdsong and ambient river sounds.

A plane lands at London City Airport. Communities in the area have used web-based noise maps to demonstrate their objections to proposed flight expansion.

Their data was then visualized in various formats to help advance their argument. Royal Docks' residents are continually plagued by planes taking-off and landing at London City Airport, and plans to expand the number of flights a year by 50% (up to 120,000) were opposed by local communities on the basis that it would decrease their quality of life.

GIS and citizen science go hand-in-hand. Web-based mapping platforms serve as useful tools for national conservation societies, local community groups and planning departments to compile tangible data on environmental issues. Voluntary, grassroots approaches can help compile lay knowledges that feed back into more formal political frameworks.

Environmental Justice

At a local level, GIS has been frequently used to engage stakeholders in the planning of environmentally 'bad' sites. Nuclear power stations, wind farms, landfill sites, and other energy facilities are often subject to NIMBY opposition for aesthetic, health and social reasons. This is despite of their capacity to produce 'good' economic factors or employment opportunities. GIS has thus found itself deployed alongside broader cost-benefit analysis (CBA), and multi-criteria decision analysis (MCDA) approaches to socio-political conflict. Environmental Justice (EJ) activists believe these decisions act to further embed racial and class divides. GIS provides an important angle to the EJ movement.

Elements of Justice

Broadly, the EJ movement is a loose connection of social groups, stakeholders and activists who have sought to contest socio-political injustices. Commonly, this has been through a single motive; the equal distribution of environmental goods and bads. As Schlosberg contends, 'the issue of distribution is always present and always key' to the guiding EJ ethos. Yet, other demands are frequently put forward. Following Schlosberg, there are two further demands that constitute the EJ movement than a mere '[re-]distribution of environmental ills and benefits. The first is the 'recognition of the diversity of participants and experiences in affected communities'. Thus, EJ demands that people affected by environmental injustices are appropriately noticed by others. A lack of recognition in local community discourses, 'demonstrated by various forms of insults, degradation, and devaluation...', marginalize those already least able to contest political decisions. The second is the notion of *participatory justice*. According to Schlosberg;'[i]f you are not recognised, you do not participate.' Thus, recognitional justice leads directly to participatory justice. In EJ terms, participation is about involving those outside the typical political/institutional order. Democratic and participatory decision-making procedures are both an element of, and a condition for, social justice. Simultaneously, institutionalised exclusion, social cultures of misrecognition, and current distributional patterns can be challenged.

An aerial photograph of New Orleans after Hurricane Katrina in 2005. The High magitude flood event disproportionately affected the city's poor, black neighbourhoods.

Flood Hazards, Race, & Environmental Justice in New York

Maantay and Maroko's research is intended to help hazard management and disaster planning before and after a high magnitude flood event. By using a Cadastral-based Expert Dasymetric System (CEDS) they were able to estimate the number of 'vulnerable sub-populations' in the densely populated New York City area. Their research broadly supports an EJ approach to natural disaster mitigation. By highlighting the importance of equality issues and the disproportionate exposure some people have to such events, they invoke the EJ movement's notion of distributional justice. The lack of 'strong social, financial, or political support structures' are constituent factors in how people deal with large-scale disasters. The criticism of the US government's response to the devastating effects of Hurricane Katrina in 2005, draw on such notions. The online publication set-up by the Social Science Research Council (SSRC) entitled Understanding

Katrina helps to ground this research in socio-political approaches to so-called 'natural' disasters.

Maantay and Maroko believe that GIS can have an important role in these 'risk-framed' understandings. As New York is a 'hyper-heterogeneous urban area', traditional administrative population data is insufficient for flood zone/population risk research. A more fine-tuned analysis is possible if tax-level datasets (based on smaller, residential units) are used instead. Maantay and Maroko use Federal Emergency Management Agency (FEMA) floodplain and tax-level datasets to determine the potential number of people at risk. A CEDS approach significantly increases the number of affected people within New York City. In using a different GIS method, Maantay and Maroko are able to better represent the impact of such flood events on minority populations. As such, their research supports all of Schlosberg's notions of distributional, recognitional and participatory justice. As noted by themselves; 'the disadvantages suffered by racial and ethnic minority communities during and after disasters are due primarily to their low economic status and lack of political power'. Their research broadly supports the aims of the EJ movement.

New Urban Landscapes

GIS has also had a role in formulating new urban landscapes. Planned cities - designed entirely from scratch - routinely use digital technologies to visualize and demonstrate urban layouts, building structures and transport arrangements. Although CAD/CAM technologies are often used to assist in the visualization, construction, and delivery of certain engineering features, GIS helps to realize distinctly spatial components of the city. Environmental narratives of a 'carbon-free' and sustainable future favour those in the GIS industry. 'The challenge of the 21st century' as ESRI would have it, 'is to arrest the progress of climate change'. Geospatial software has played its part in developing this narrative.

Masdar City

Masdar City is a 'sustainable, zero-carbon, zero-waste' project currently under construction in the United Arab Emirates (UAE). Situated in the Abu Dhabi emirate, Masdar is described as 'an emerging global hub for renewable energy and clean technologies'. The Abu Dhabi Future Energy Company have funded and overseen the $18bn dollar project. No cars are allowed on its streets, energy is produced in part by renewable sources, building materials are 'sustainable' and water-use is controlled. GIS is being employed to plan, facilitate and test a plethora of environmental phenomena and technological processes.

A dedicated GIS team is responsible for 'managing the overall spatial information needs' of the project, starting with the drawing of a common base map with which to support the city's infrastructure. Without a spatial plan of Masdar's operative mechanisms, the city will fail to deliver its grand ambition. In particular, GIS is being used to visualize, analyse and model *land-use* in the city. Masdar – unlike any other city – has to incorporate a wealth of energy-related facilities within its perimeter. As EJ activists are all too aware, the siting of such facilities can be a key area of conflict. Masdar's water treatment and sewage plants, material recycling centre, solar power plant, geothermal test site, solar panel test field and concrete batching plant all need to be situated inside the city's boundaries. As CH2M HILL's Site Control and GIS Manager for the Masdar project confirms; 'never have so many environmental facilities come together in one place'. GIS is the central tool with which to imagine – in a digital environment – different siting scenarios. In this case, GIS is seen to operate as a decision-making tool; informing the practitioners who work on the Masdar project.

GIS is also being used to model some of Masdar's key infrastructural features directly. Its involvement in simulating the city-wide Personal Rapid Transport System (PRTS) is one such example. As common road vehicles are banned from the city, the 'driverless' transit system will transport people and freight across the 7km2 area. GIS is capable of modelling the system route, due to comprise 85 passenger stations and approximately 1,700 automated vehicles. By drawing spatial buffer zones around potential PRTS stops, passenger-distance maps can visualize residential areas that fall outside of ideal service requirements. GIS is an instrumental tool in visualizing such problems. A smooth, functioning PRTS is a central infrastructural aspect of Masdar's grand vision, and engineering companies who specialize in GIS technologies have helped in realizing this digitally-orchestrated dream. Yet journalists in particular have been sceptical. As Bryan Walsh has argued; [wi]ll Masdar City ever really develop the authenticity of a real city?'. Or as Jonathan Glancey contends, will its 'ultra-modern aspects...prove to be a mirage'?

Post-political Agendas

'Post-politics' is a neologism for the consensual, participatory and techno-managerial approach to modern governance. Originally coined by Slavoj Zizek, and discussed by the likes of Jacques Rancière and Erik Swyngedouw, the post-political critique argues that life in the Western world is routinely characterized by the de-politicizing effects of a 'consensual police order'. A number of different techno-managerial 'fixes' have been sought by neoliberal governments in order to solve expressly environmental problems, rather than due political processes. As Swyngedouw has argued, such forces have 'replaced debate, disagreement and dissensus with a series of technologies of governing that fuse around consensus, agreement, accountancy merits and technocratic environmental management'. Thus if the rise of the post-political order is due to the increasing reliance upon 'technocratic environmental management', as Swyngedouw has argued, then GIS – as a tool for neoliberal environmental governance – is implicit in such an order.

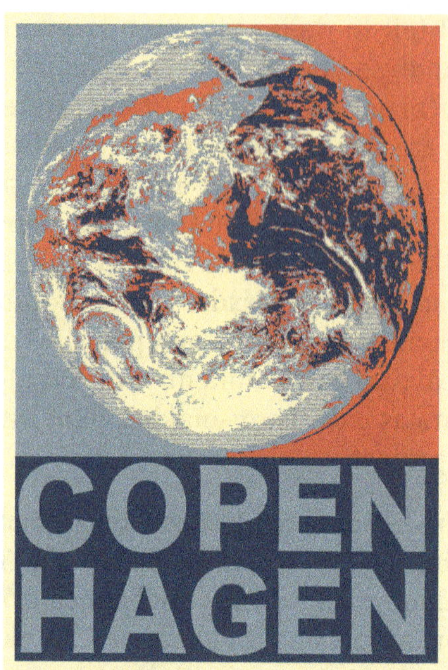

The 2009 Copenhagen Summit failed to produce a legally binding treaty on national CO2 levels, despite (or perhaps because of) a near-global delegation. An example of Jacques Ranciere's partition of the sensible in action. IPCC data, based on GIS and environmental modelling processes, provide the basis for these proposed legal frameworks.

The Police, the Political and Politics

Firstly, it serves to elucidate upon claims to a 'post-political impasse'. This is best understood through what Ranciere calls *the police, the political* and *politics*. It is within these three terms that Ranciere carves out what he calls the true meaning of political action, and of what it is to exercise political right. As Panagia neatly summarizes; 'politics is the practice of asserting one's position that ruptures the logic of arche'. Politics is about rediscovering the art of debate, conflict and struggle, and not merely about re-organizing the administrative framework of existing political structures (namely, the state apparatus). *The political* – in Ranciere's words – is for '[t]he one who is 'unaccounted-for', the one who has no speech to be heard'. Democracy does not work towards an 'idealized-normative condition' of equal rights, but is built upon the very ontological notion of such. *Politics* brings the political into the foreground; rendering that which was previously noise, legitimate speech. Much of what is thought to be politics in the contemporary world is actually subsumed within *the police.* Policy is not politics is this sense, then. This is what Ranciere calls *the partition of the sensible*, or the established order of things.

The post-politics of contemporary governance brings all that it can into this order. Those who were previously cast outside the police structure are now "responsible' partners' in a stakeholder-based arrangement. All views that were previously antagonistic and conflictual are now brought together in a more homely, consensual arena. No room is made for 'irrational' demands. As such, Ranciere says that, '[c]onsensus is the reduction of politics to the police'. Nowhere has this been more visible than within the apocalyptic climate narratives told *ad infinitum* by environmental practitioners, policy-makers and non-governmental organizations. Consensual politics has found its home in the environmental arena.

Post-political GIS?

In short, GIS works as a tool for mediating and diffusing socio-environmental conflicts. It does so by working within Ranciere's notion of the partition of the sensible. Whilst it may allow previously unheard voices to gain a voice (in environmental justice campaigns, foremostly), it still does not – as a tool of neoliberal governance – make room for those who are deemed 'outside', unruly or conflictual to have a voice. For example, Elwood laments the notionally 'participatory' flag-waving carried out by those involved in urban GIS-based projects. As she says; 'the skills and financial and temporal costs of using GIS effectively bar many individuals, social groups and organizations from participation in research and decision-making where it is used', denying those without the means to participate, from participation. GIS does not necessaily facilitate involvement for all.

Moreover, GIS is limited in its ontological scope, reducing all things spatial to a *calculable order*. As Leszczynski contends, GIS operates a 'discourse populated by discrete objects of knowledge'; differentiating 'between the binary of truth and error'. GIS is thus a central *polic-ing* tool for contemporary socio-environmental governance. It works to order space into discrete and ordered formats. As Jeremy Crampton points out, 'the basic model of the world in GIScience texts is: points, lines, areas, surfaces and volumes', ill prepared to deal with non-discrete, continuous environmental phenomena. All that does not fit into this order is left aside. GIS thus excludes that which cannot be 'structured accordingly', excluding epistemologies that do not neatly fit into the formal computing framework; a formal framework synonymous with 'the accountancy calculus of risk and the technologies of expert administration'. The apparatus of choice for the techno-enviro-managerialist, if Ranciere and Swyngedouw's analyses are to be put forward, and Elwood, Crampton and Leszczynski's criticism's are to be accepted, is GIS.

References

- Young, Iris Marion (1990). Justice and the politics of difference (11th print. ed.). Princeton, N.J.: Princeton University Press. ISBN 0-691-07832-7.
- McElvaney, Shannon. "Masdar City Development Programme: Using GIS Technologies to Help Plan and Build a Sustainable City" (PDF). Retrieved 16 May 2011.
- Glancey, Jonathan (10 May 2011). "Inside Masdar City: a modern mirage". London: The Guardian. Retrieved 17 May 2011.
- Panagia, Davide (2001). "Ceci n'est pas un argument: An Introduction to the Ten Theses". Theory

and Event. 5 (3). Retrieved 16 May 2011.

- Crampton, J. W. (2011). "Cartographic calculations of territory". Progress in Human Geography. 35 (1): 92–103. doi:10.1177/0309132509358474.
- Goodchild, Michael F. (July 2009). "GIScience and Systems". International Encyclopedia of Human Geography: 526–538. doi:10.1016/B978-008044910-4.00029-8. Retrieved 15 May 2011.
- Audubon. "Audubon Science: Christmas Bird Count" (PDF). National Audubon Society. Retrieved 16 May 2011.

7

Geographic Information Systems: A Historical Perspective

The geographic information system that stores and analyses data of the past geographies is referred to as historical geographic information system. The following content mentions topics such as Great Britain historical GIS and HistoAtlas. The chapter serves as a source to understand the historical perspective of geographic information system.

Historical Geographic Information System

A historical geographic information system (also written as historical GIS or HGIS) is a geographic information system that may display, store and analyze data of past geographies and track changes in time. It can be regarded as a subfield of historical geography and geographic information science.

GIS was originally developed for use in environmental sciences, military and for computer assisted cartography. It is the opinion of some that the tools developed for these uses are ill suited for the features of historical data.

Techniques used in HGIS

- Digitization and georeferencing of historical maps. Old maps may contain valuable information about the past. By adding coordinates to such maps, they may be added as a feature layer to modern GIS data. This facilitates comparison of different map layers showing the geography at different times. The maps may be further enhanced by techniques such as rubbersheeting, which spatially warps the data to fit with more accurate modern maps.
- Reconstruction of past boundaries. By creating polygons of former administrative sub-divisions and borders, aggregate statistics can be compared through time.
- Georeferencing of historical microdata (such as census or parish records). This enables the use of spatial analysis to historical data.

Notable Historical GIS projects

- HistoricalGIS.com, Longitudinal Urban Historical GIS Projects in the Canadian

Cities of London, Victoria, Montreal, and Windsor. Housed at the Human Environments Analysis Laboratory, University of Western Ontario. Public access via HistoricalGIS in Canada.

- HistoricalGIS.org, A site dedicated to the study and promotion of HGIS. Includes Projects, Lists, and other resources for students, educators, researchers, and the general public. Public access via HistoricalGIS.org.
- Great Britain Historical GIS, A GIS enabled database holding diverse geo-referenced maps, statistics, gazetteers and travel writing, especially for the period 1801-2001 covered by British censuses. Public access via the Vision of Britain site. Created and maintained by Portsmouth University.
- China Historical GIS similar project for Imperial China developed by the universities of Harvard and Fudan, China.
- David Rumsey Historical Map Collection, one of the world's largest map collections, which has digitized and georeferenced a large part of its collection and published it on the internet.
- Electronic Cultural Atlas Initiative (ECAI) a clearinghouse for the exchange of metadata of Historical GIS. Maintained by the University of California, Berkeley.
- HGIS Germany Institute of European History (Mainz) and Institute i3mainz at the University of Applied Sciences
- HisGIS Netherlands includes vectorized real estate boundaries of the oldest cadastral maps of several regions of the Netherlands, including Amsterdam, which have been linked to historical registers such as election, tax revenue and parish registers. Developed by the Fryske Akademy
- Belgian Historical GIS tracks the development of administrative boundaries in Belgium since 1800. Developed by the University of Ghent
- The National Historical Geographic Information System (NHGIS) system for displaying and analyzing Census tracts and tract changes in the United States.
- HistoAtlas is an open historical geographical information system that tries to build a free historical atlas of the world.
- Atlas-Historical Cartography This website provides information on the evolution of administrative boundaries of Portugal, and on censuses and other statistical series for the nineteenth and twentieth centuries. Public access via the Atlas-Historical Cartography.
- Mammoth Cave Historical GIS documents the people who lived in the Mammoth Cave region before it became a national park. Public access via MCHGIS.

- Bibliosof-HGIS Russian National Historical Geographic Information System .
- Wikimaps project that will integrate georeferences maps on Wikimedia Commons, Wikidata and Wikipedia

Software or Web services Developed for Historical GIS

- TimeMap — A Java open-source applet (or program) for browsing spatial-temporal data and ECAI data sets Developed by the department of archaeology University of Sydney.
- Version 4+ of Google Earth added a time line feature that enables simple temporal browsing of spatial data

Great Britain Historical GIS

The Great Britain Historical GIS (or GBHGIS), is a spatially enabled database that documents and visualises the changing human geography of the British Isles, although is primarily focussed on the subdivisions of the United Kingdom mainly over the 200 years since the first census in 1801. The project is currently based at the University of Portsmouth, and is the provider of the website *A Vision of Britain through Time.*

NB: A "GIS" is a Geographic Information System, which combines map information with statistical data to produce a visual picture of the iterations or popularity of a particular set of statistics, overlaid on a map of the geographic area of interest.

Original GB Historical GIS (1994-99)

The first version of the GB Historical GIS was developed at Queen Mary, University of London between 1994 and 1999, although it was originally conceived simply as a mapping extension to the existing Labour Markets Database (LMDB). The system included digital boundaries for Registration Districts and Poor Law Unions (c.1840 to 1911), Local Government Districts (1911 to 1974), and Parishes (1870s to 1974). These boundaries were held not as polygons but as line segments (*arcs*), using ArcGIS software. Dates of creation and abolition were held for each line segment (or "arc") and custom software was developed to assemble line segments into polygons, creating conventional boundary maps for particular dates. Meanwhile, the Labour Markets Database evolved into the Great Britain Historical Database (GBHDB), which stored a large collection of historical statistics from the census, vital registration and records of poverty and economic distress. These were held in thousands of columns within hundreds of separate tables, within an Oracle database. This system is described in detail in Gregory and Southall (1998), and in Gregory and Southall (2002).

New GB Historical GIS (2000-)

The second version of the GB Historical GIS was developed at the University of Portsmouth from 2000 onwards. The work was mainly funded by the UK National Lottery, so the results had to be useful to a far wider audience than most historical GIS projects.

New Architecture

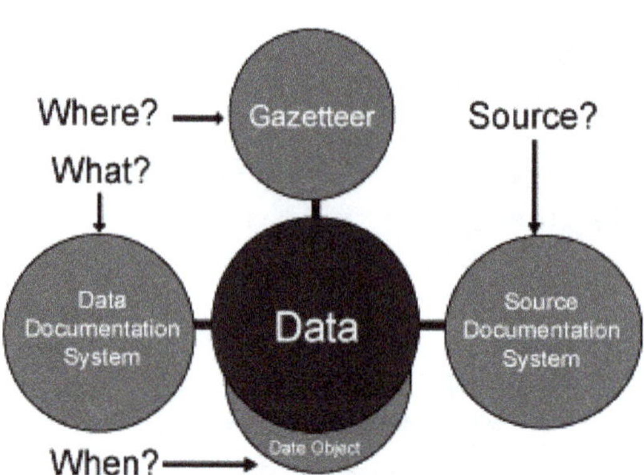

This is a true spatial database in which all content is held in Oracle, although GIS software is used to edit content. It is designed to overcome limitations of the original system:

- The statistical content is now the core of the system, all data values being held in a single column of a single table, with other columns indicating what the number measures, when and where it is for, and the source it was taken from (Southall, 2007).
- Where is recorded not directly as a location but via a reference to a large catalogue of administrative units. This catalogue is organised as an ontology, each unit having any number of names and at least one *IsPartOf* relationship with a higher-level unit; the obvious exception is the root unit, which represents the British Isles and to which all other units ultimately belong.
- Almost all the original digital boundaries are included in the new system, but they are held as polygons rather than line segments. Many units, especially those lacking associated statistical data, do not have boundary polygons. Most of these have approximate centroids, inferred from their relationships with units that do have polygons.
- The meaning of the statistical content—the what in the central data table—is recorded via a data documentation subsystem which is another ontology. It was designed as a relational implementation of the aggregate data extension developed by the Data Documentation Initiative. This sub-system does not simply provide text defining variables, it directly drives the graphical presentation of data. Each data value is located within an nCube or Hypercube.

Expanded Content

This new version of the GB Historical GIS also included several other kinds of content:

- Descriptive Gazetteers: Over 90,000 entries from three late nineteenth century gazetteers: John Marius Wilson's *Imperial Gazetteer of England and Wales* (1872); Frances Groome's The Ordnance Gazetteer of Scotland (1885); and John Bartholomew's Gazetteer of the British Isles (1887).
- Travel Writing: The text of most of the best known historical British travel writers, including James Boswell, William Camden, William Cobbett, Daniel Defoe, Celia Fiennes, Charles Wesley and Arthur Young. The earliest source included in the GB Historical GIS is a survey of Wales written by Giraldus Cambrensis in 1188. Place-names are identified within these texts using XML tags defined by the Text Encoding Initiative. This is believed to be the largest collection of British historical travel literature on the web, and is unique in that it is fully geo-referenced.
- Census Reports: The sub-system recording sources of statistical information holds a complete list of all the tables published in British census reports up to 1961, enabling the system to reconstruct selected tables. The system also holds the introductory text from selected reports, and the *Guide to Census Reports: Great Britain 1801-1966.*
- Geographical name authority: the administrative unit ontology described above was created from quite separate sources from the original GIS, including Frederick Youngs' *Guide to the Local Administrative Units of England*, Melville Richards' *Welsh Administrative and Territorial Units* and a new gazetteer of Scottish counties, parishes and burghs created by the Scottish Archives Network. It also holds additional variant names found in census reports, and is designed to be used for name authority control.

All of this new content is held in the same Oracle database and linked to the polygons and statistics inherited from the original GBHGIS.

Historical Maps

A GIS consisting entirely of administrative boundaries can create maps but these are hard to relate to the real world. The project has therefore constructed a second GIS consisting entirely of scanned images of historical maps, supporting an on-line map library.

Three complete sets of one inch to one mile maps of Great Britain have been scanned and geo-referenced, each accompanied by less detailed maps from the same period:

- The Ordnance Survey New Popular Edition, from the late 1940s. These are the most recent detailed maps of Britain to be free from OS copyright. The smallest scale twentieth century map is *New Map of the British Isles. Produced under the direction of A. Gross*, (London: Geographia, 1921; British Library shelfmark Maps 1080.(70.)). The intermediate mapping is the *Ordnance Survey of Great Britain. Scale of ten statute miles to one inch. 1:633 600 maps from 1904* (British Library shelfmark Maps 1125.(14.)).

- The Ordnance Survey First Series. These were created over several decades during the mid-19th century, and the GB Historical GIS uses the earliest *state* for each sheet held by the British Library. The least detailed nineteenth century map is from 1812 and is by Robert Wilkinson, at a scale of 1:1,625,000 (British Library shelfmark Maps 177.d.2.(15.)). The intermediate scale map is Smith's *New Map of the United Kingdom of Great Britain and Ireland: on which the Turnpike, and Principal Cross Roads, are carefully described. Particularly distinguishing the Route of the mail Coaches, the course of the Rivers, and Navigable canals; ...*, published in 1806 at a scale of 1:633,600 (British Library shelfmark Maps 177.d.2.(14.)).

- The Land Utilisation Survey of Great Britain, created by L.Dudley Stamp of the London School of Economics. These maps include all the published one inch sheets, plus the 56 maps covering upland Scotland, hand painted in water colour to show land use, that Stamp deposited with the Royal Geographical Society (RGS Control No. 568206). The ten mile to the inch summary sheets published by the LUSGB are also included.

This collection of historical maps is not held in the main Oracle system. They are instead managed using open source MapServer software. However, they are mainly accessed via MapServer's implementation of the Open Geospatial Consortium's Web Map Service standard. This is how they are used by the GB Historical GIS project's Vision of Britain system, but they are also available for use as base maps by other web sites.

Re-districting Statistics to Constant Units

Britain has had an unusually large number of changes to its local government geography, and the current districts date back only to 1996, to 1974 or, in London, to 1965. As census reporting has always been based on local government units, it is hard to study how any particular area has changed in the long term. One of the main reasons for building the GB Historical GIS was to enable demographic and social statistics to be '*re-districted* from various historical units to modern districts. This is done using a vector overlay methodology, using parish-level counts of total population to weight the reallocation of district-level data.

This methodology has been used to replicate the most important statistics from the Key Statistics release from the 2001 census for many earlier dates, including total population from 1801, occupational structure for selected censuses from 1841 onwards, and age and gender structure for every census from 1851 onwards.

A Vision of Britain through Time

Components of the GB Historical GIS are available for download by academic researchers from the UK Data Archive and from EDINA's UKBORDERS system. However, the main way most people can access the system is via the Vision of Britain web site, developed by the GB Historical GIS project with their lottery funding.

The site is designed mainly as a resource for studying local history but also includes extensive mapping facilities. It includes home pages both for "places", i.e. towns and villages, and for the individual administrative units based on places. Administrative unit pages provide access to census statistics for the unit, to boundary maps and to formal information on official names and status, relationships with other units and boundary changes. All these web pages are generated by software from the data held in the underlying GB Historical GIS. Many Wikipedia pages refer to Vision of Britain.

Vision of Britain is an unusual web site as it is database-driven, but uses the ontologies in the underlying system to create clickable links between pages: most pages the site can create can be reached without filling out a search form, or clicking on an image map, and this makes the site's content generally accessible to search engines. One result is that Google searches for historical information for particular places in Britain are very likely to return links to Vision of Britain. For the most reliable results, search in Google for "place county history"; for example, "Portsmouth Hampshire History".

For a detailed guide to using the Vision of Britain system for research into local history.

Is it a GIS at all?

The post-2000 GB Historical GIS makes no use of commercial GIS software, except for editing parts of the content, and implements a data model which could not be implemented using packages such as ArcGIS or MapInfo, so is it a GIS at all? It is certainly not a conventional GIS, but one answer is that any system that can create an image like the one shown below is some kind of GIS. This image from *A Vision of Britain through Time* combines the boundaries of local government districts, data on unemployment from the 1931 census, and a scanned image of an Ordnance Survey ten mile-to-one inch map from the early 20th century.

Extended Historical GIS (2007 Onwards)

A new system is being developed, partly with funding from the European Union under the QVIZ project, which will no longer be limited to Great Britain:

- The *root unit* represents the world rather than the British Isles, although more detailed decisions about map projections mean that the system is in practice limited to Europe.
- All coordinates are held using latitude and longitude, not the Ordnance Survey National Grid.
- All geographical names and some other text are held using Unicode (UTF-8).
- Multiple languages are supported, especially when recording geographical names, using Ethnologue codes to identify modern languages and Linguist codes for historical ones.
- An enhanced web site based on this extended system was launched in 2009.

HistoAtlas

HistoAtlas is a free collection of historic geographic information of the human culture all over the world. This is achieved as a time enabled geographic information system (GIS) on the web. All information can be used and edited freely and is intended to be a resource for education, archaeologists, historians and others.

Introduction

HistoAtlas provides a system to actively maintain historical information derived from historical records and check if it is consistent. It is not meant for the discovery of new historical facts but to put everything together so it can be presented as one whole story.

It is an open project making sure everyone benefits from it. Everyone can use the information and can collaborate on the project. Its main audience is the general public but it should also have enough historical details so also historians should be able to enjoy it.

HistoAtlas is not only able to visualize the changes in extent of different countries, but also the events that caused this change altogether, because these things are historically more important than just the change of a border.

The atlas has information about different aspects of history. A few examples.

- Political boundaries based on antique historical records or archeological research.
- Historical events like wars, disasters, discoveries, treaties and journeys that shaped the course of time.
- Facts about historical figures and their families that played an important role in history.
- The evolution of cultural aspects like languages and religions.

Strategy

HistoAtlas aims to be a free, multilingual historical encyclopedia, intended as the most precise open content global historical reference.

In concept there are some similarities to what online encyclopedia like Wikipedia are doing, with differences on how data is stored, presented and used. It is a history oriented geographic information system. Information is structured more transparently so it can be searched more efficiently and presented in different ways. Wikipedia and other encyclopedia are focused on articles and are not able to create maps efficiently because they are not meant to do this.

But like Wikipedia the information and application will be made available under an open license, meaning it will be maintained by volunteers who want to share their knowledge.

Consistency

The vision that HistoAtlas wants to put forward is the one of a collaborative system. Scientists require that the information is correct. Making sure that data is correct will be a high priority of the system. For now only a basic system has been put in place, but this will be extended in the future.

Licensing

Everything developed for the project is licensed under an open license. It can be used and improved by anyone.

All data published under the project is put under a Creative Commons Attribution-ShareAlike License.

Permissions

All chapters in this book are published with permission under the Creative Commons Attribution Share Alike License or equivalent. Every chapter published in this book has been scrutinized by our experts. Their significance has been extensively debated. The topics covered herein carry significant information for a comprehensive understanding. They may even be implemented as practical applications or may be referred to as a beginning point for further studies.

We would like to thank the editorial team for lending their expertise to make the book truly unique. They have played a crucial role in the development of this book. Without their invaluable contributions this book wouldn't have been possible. They have made vital efforts to compile up to date information on the varied aspects of this subject to make this book a valuable addition to the collection of many professionals and students.

This book was conceptualized with the vision of imparting up-to-date and integrated information in this field. To ensure the same, a matchless editorial board was set up. Every individual on the board went through rigorous rounds of assessment to prove their worth. After which they invested a large part of their time researching and compiling the most relevant data for our readers.

The editorial board has been involved in producing this book since its inception. They have spent rigorous hours researching and exploring the diverse topics which have resulted in the successful publishing of this book. They have passed on their knowledge of decades through this book. To expedite this challenging task, the publisher supported the team at every step. A small team of assistant editors was also appointed to further simplify the editing procedure and attain best results for the readers.

Apart from the editorial board, the designing team has also invested a significant amount of their time in understanding the subject and creating the most relevant covers. They scrutinized every image to scout for the most suitable representation of the subject and create an appropriate cover for the book.

The publishing team has been an ardent support to the editorial, designing and production team. Their endless efforts to recruit the best for this project, has resulted in the accomplishment of this book. They are a veteran in the field of academics and their pool of knowledge is as vast as their experience in printing. Their expertise and guidance has proved useful at every step. Their uncompromising quality standards have made this book an exceptional effort. Their encouragement from time to time has been an inspiration for everyone.

The publisher and the editorial board hope that this book will prove to be a valuable piece of knowledge for students, practitioners and scholars across the globe.

Index

www.ingramcontent.com/pod-product-compliance
Lightning Source LLC
Chambersburg PA
CBHW061934190326
41458CB00009B/2734
9781635491333